Daily Skill-Builders

Science & Technology

Grades 4–5

Writer
Sasha Rancourt-Thomas

Editorial Director
Susan A. Blair

Project Editor
Holly Moirs

Cover Designer
Roman Laszok

Interior Designer
Mark Sayer

Production Editor
Maggie Jones

WALCH PUBLISHING

1 2 3 4 5 6 7 8 9 10
ISBN 0-8251-5099-X

Walch Publishing
P. O. Box 658 • Portland, Maine 04104-0658
walch.com
Printed in the United States of America

WALCH PUBLISHING

Table of Contents

Daily Skill-Builders

Science & Technology

Grades 4–5

To the Teacher . . . . *v*

SCIENCE AS INQUIRY 1–13

PHYSICAL SCIENCE 14–47

LIFE SCIENCE 48–89

EARTH SCIENCE 90–140

SCIENCE AND TECHNOLOGY 141–146

PERSONAL AND SOCIAL PERSPECTIVES 147–161

THE HISTORY AND NATURE OF SCIENCE 162–180

ANSWER KEY 181

To the Teacher

Introduction to *Daily Skill-Builders*

The *Daily Skill-Builders* series began as an expansion of our popular *Daily Warm-Ups* series for grades 5–adult. Word spread, and eventually elementary teachers were asking for something similar. Just as *Daily Warm-Ups* do, *Daily Skill-Builders* turn extra classroom minutes into valuable learning time. Not only do these activities reinforce necessary skills for elementary students, they also make skill-drilling an engaging and informative process. Each book in this series contains 180 reproducible activities—one for each day of the school year!

How to Use *Daily Skill-Builders*

Daily Skill-Builders are easy to use—simply photocopy the day's activity and distribute it. Each page is designed to take approximately ten to fifteen minutes. Many teachers choose to use them in the morning when students are arriving at school or in the afternoon before students leave for the day. They are also a great way to switch gears from one subject to another. No matter how you choose to use them, extra classroom minutes will never go unused again.

Building Skills for All Students

The *Daily Skill-Builders* activities give you great flexibility. The activities can be used effectively in a variety of ways to help all your students develop important skills, regardless of their level.

Depending on the needs of your students and your curriculum goals, you may want the entire class to do the same skill-builder, or you may select specific activities for different students. There are several activities for each topic covered in *Daily Skill-Builders,* so you can decide which and how many activities to use to help students to master a particular skill.

If a student does not complete an activity in the allotted time, he or she may complete it as homework, or you may allow more time the next day to finish. If a student completes a skill-builder early, you may want to assign another. *Daily Skill-Builders* give you options that work for you.

Students in one grade level vary in their abilities, so each *Daily Skill-Builders* covers two grades. In a fourth-grade class, for example, some students may need the books for grades 3–4. Other students may need the greater challenge presented in the 4–5 books. Since all the books look virtually the same and many of the activities are similar, the students need not know that they are working at different levels.

No matter how you choose to use them, *Daily Skill-Builders* will enhance your teaching. They are easy for you to use, and your students will approach them positively as they practice needed skills.

Name ____

Classmate Clues

The information below was collected in a classroom survey. Read the information, and answer the questions that follow.

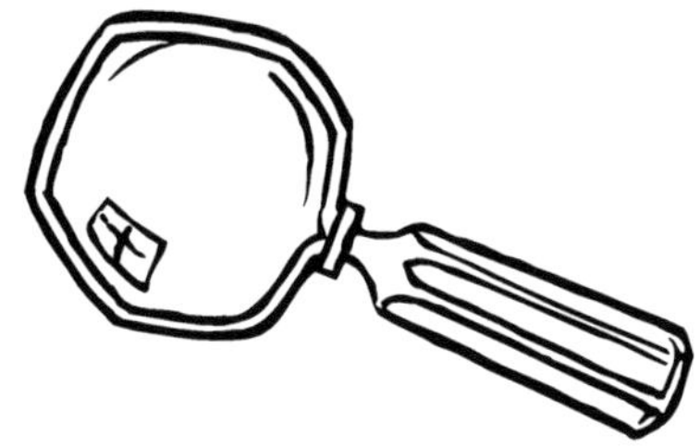

Number of students who have blue eyes: 5

Number of students who have green eyes: 3

Number of students who have brown eyes: 6

Number of students who have hazel eyes: 2

1. Organize the information that was collected in the table below.

2. What is the dominant eye color in the class? ______________

3. Why do you think that eye color occurs most often in the class?

____

____

Name ______________________________

Survey Superstars

Design a survey to obtain information about your classmates. Use the process described below. Answer each question in a complete sentence.

1. What would you like to know about your classmates? (shoe size, favorite food, and so forth)

   ____

   ____

2. How are you going to collect the information? (ask, hand out survey, and so forth)

   ____

   ____

3. How are you going to organize this information? (table, graph, pie chart, and so forth)

   ____

   ____

4. What do you think you will find out from your survey? (Favorite food is pizza, most common shoe size is a four, and so forth.)

   ____

   ____

Name ______________________________

Conducting a Survey

Answer the questions below.

1. What is the purpose of a survey? ______________________

____

____

2. How would you collect data? Give a detailed description.

____

____

____

3. How would you organize data? Give a detailed description.

____

____

____

____

____

Name ____________________

Survey Solutions

Derek conducted a survey of his class. He wanted to know how many had blue eyes and how many had brown eyes. He also noted the hair color of each person. Read Derek's chart. Then answer the questions that follow.

Blue Eyes Blonde Hair	Blue Eyes Red Hair	Blue Eyes Brown Hair	Brown Eyes Blonde Hair	Brown Eyes Red Hair	Brown Eyes Brown Hair
4 students	0 students	2 students	2 students	1 student	6 students

1. How many students were in Derek's survey?____________________
2. How many students have blue eyes?____________________
3. How many students have brown eyes?____________________
4. How many students have blonde hair?____________________
5. How many students have red hair?____________________
6. How many students have brown hair?____________________
7. What is the most common hair color? Why do you think so?

____

____

8. What is the most common eye color? Why do you think so?

____

____

9. Is there any relationship between hair color and eye color?

____

____

Name ____

Balancing Act

A triple beam balance measures the mass of objects in grams. What is the mass shown on each of the balances below?

1.

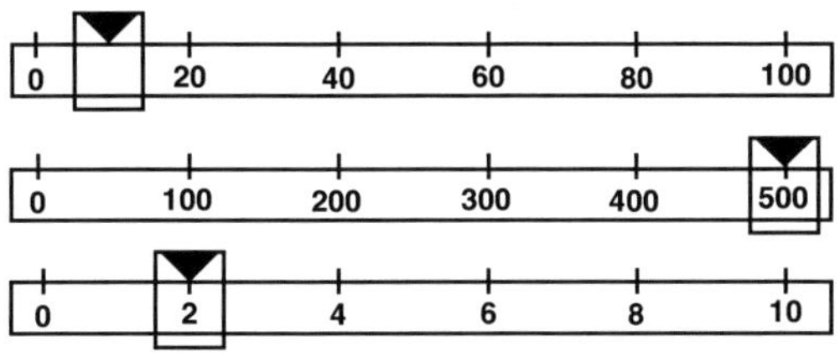

Answer________________

2.

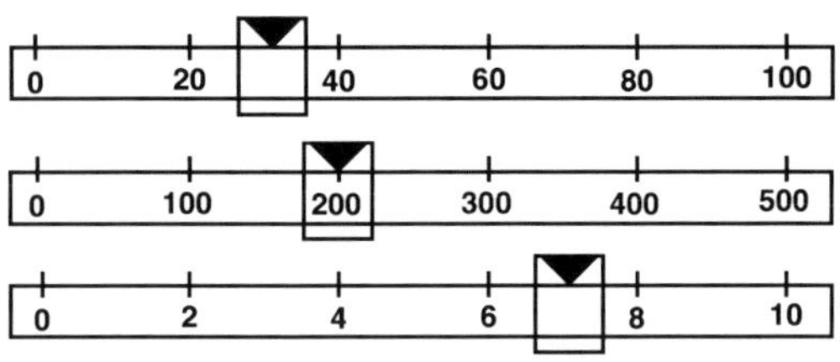

Answer________________

3.

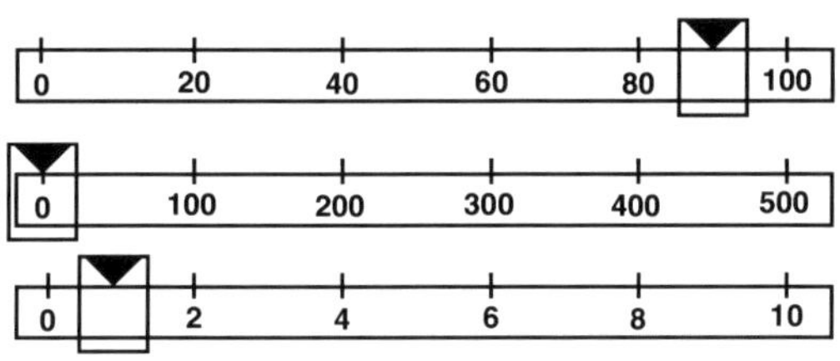

Answer________________

4.

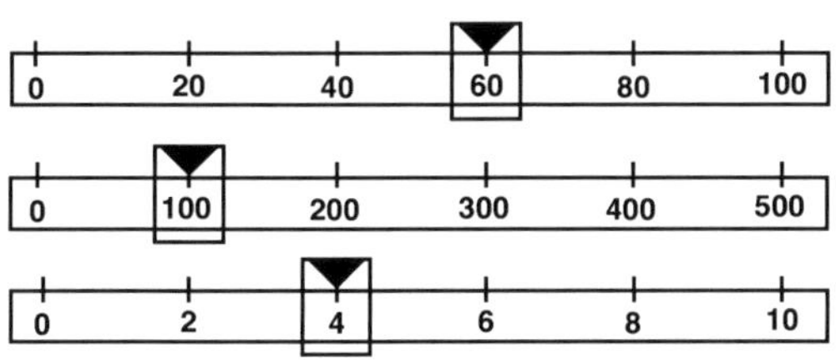

Answer________________

5.

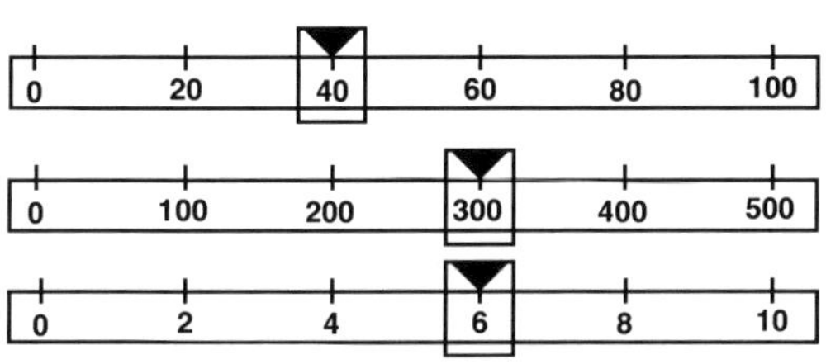

Answer________________

Name ____

Measuring the Little Things

Use a ruler that measures in centimeters to find the length of the lines below.

Measure in Centimeters:

1. __________________ Answer______________cm

2. __________ Answer______________cm

3. ________________________ Answer______________cm

4. _ Answer______________cm

5. ______________________________________ Answer____________cm

Measure in Millimeters:

6. ___ Answer____________mm

7. _____________ Answer_______________mm

8. _______ Answer_____________mm

9. _ Answer_____________mm

10. __________ Answer____________mm

Name ______________________________

Lots of Liquids

A graduated cylinder is used to measure the volume of liquids in mL. Use the graduated cylinders below to determine the volume of each.

1.

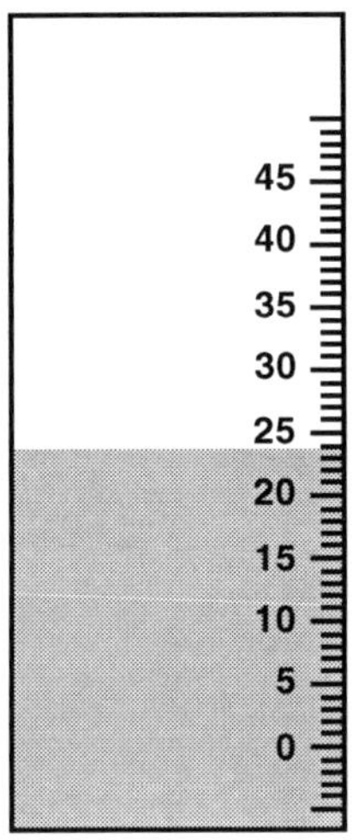

Answer __________ mL

2.

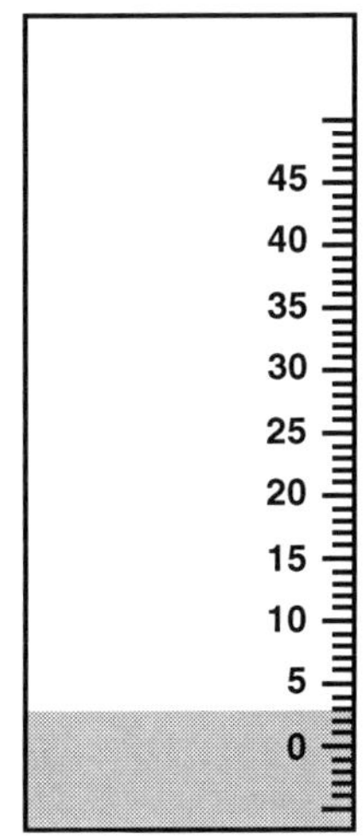

Answer __________ mL

3.

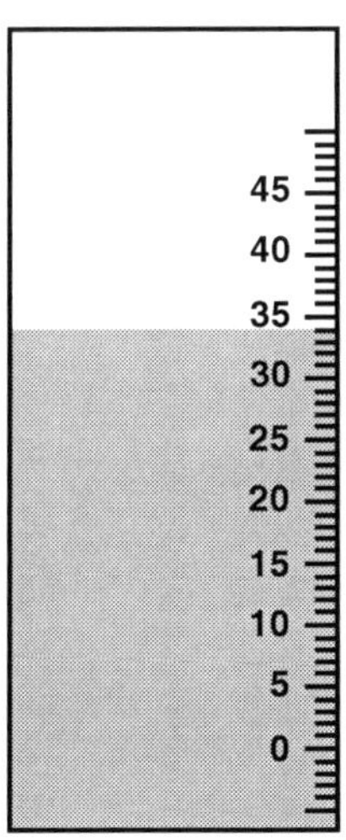

Answer __________ mL

4.

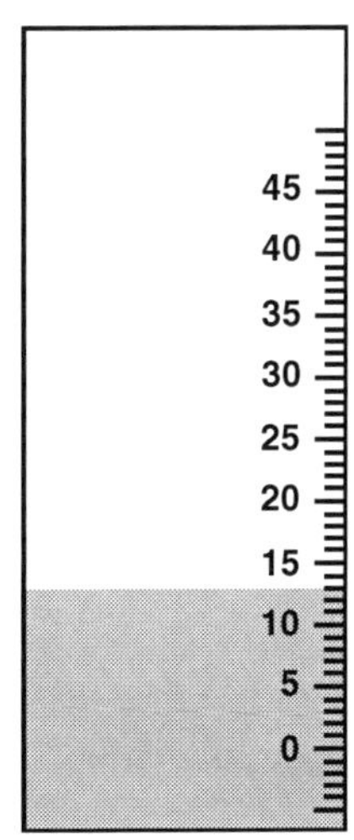

Answer __________ mL

Name ____________________

Temperature Time!

When measuring temperatures in science class you will use degrees Celsius. Practice measuring by using the thermometers below.

1.

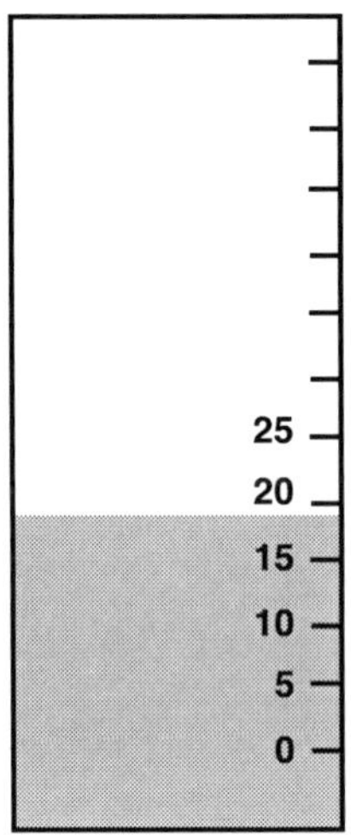

Answer ________ °C

2.

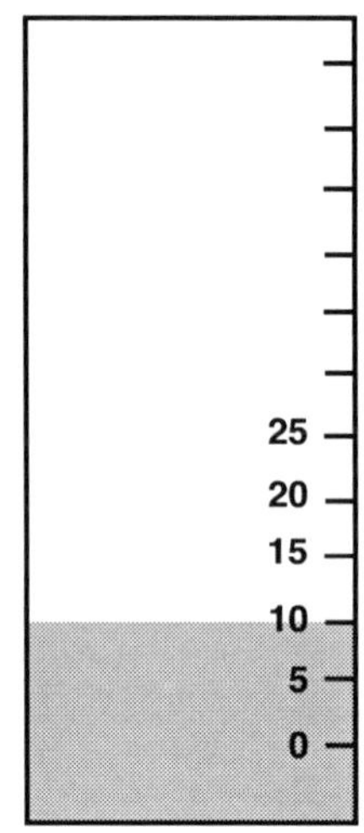

Answer ________ °C

3.

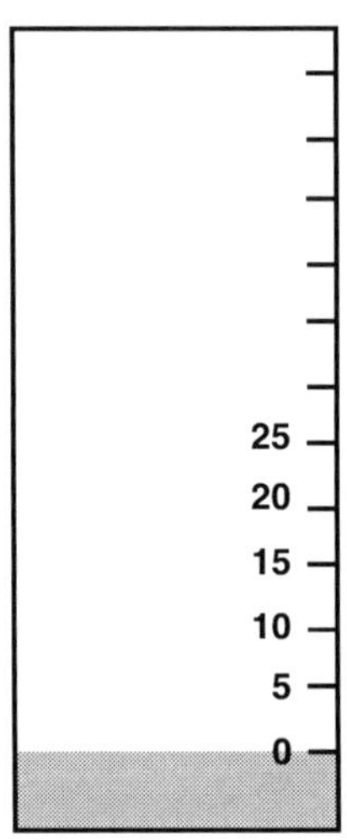

Answer ________ °C

4.

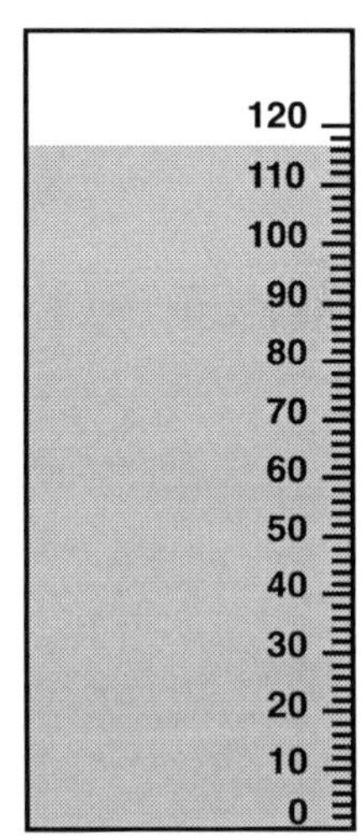

Answer ________ °C

Name ____

Quick Conversions

Use the conversions in the box to help convert the measurements below. Show all of your calculations.

10 mm = 1 cm	100 cm = 1 m	1,000 m = 1 km

1. 20 mm = _____cm

2. 400 cm = _____m

3. 4,000 m = _____km

4. 20 cm = _____mm

5. 6 km = _____m

6. 4 m = _____cm

Name ____________________

Magic Measuring

Instruments you will need: ruler, triple beam balance, graduated cylinder, thermometer

Choose the correct instrument from the box above to measure each item below. Write your answers on the lines provided.

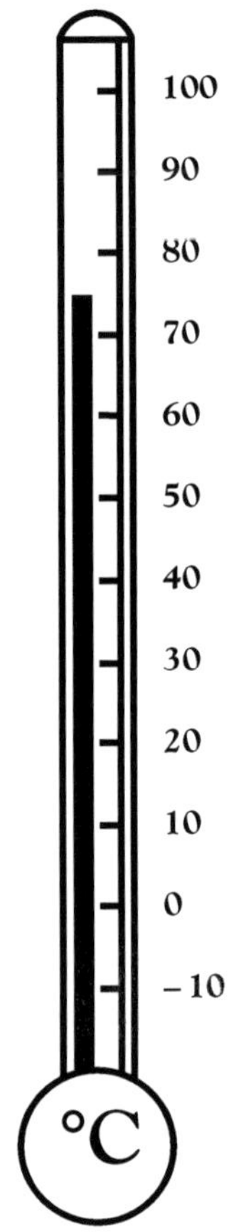

1. Measure the length of your desk in centimeters.__________
2. Measure the length of your shoe in centimeters.__________
3. Measure the diameter of a pencil eraser in millimeters.__________
4. Measure the length of a paper clip in millimeters.__________
5. Find the volume of liquid that fills your teacher's coffee mug in milliliters.__________
6. Find the mass of a tack in grams.__________
7. Find the mass of a marker in grams.__________
8. Find the temperature of the water in the school drinking fountain.__________
9. Find the temperature of tap water.__________

Name ____

Compare and Contrast

Look outside and find a tree and a flower or small plant that are growing near each other. Compare and contrast the tree and the flower or small plant using the Venn diagram below.

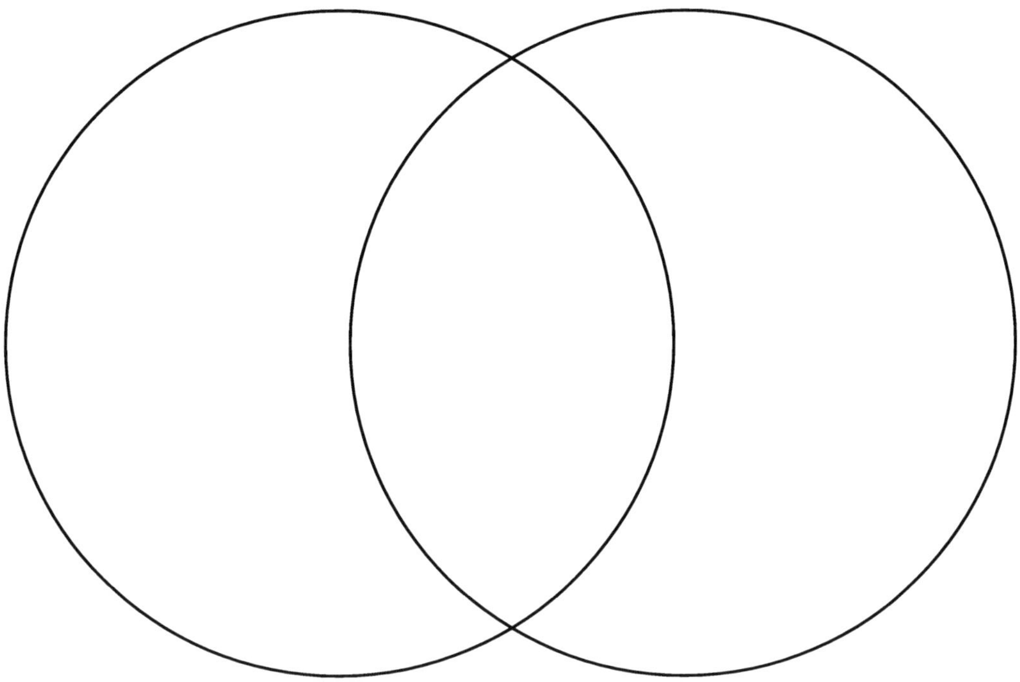

Note the similarities and differences. Then write a short paragraph comparing and contrasting the tree and the flower or small plant.

____

____

____

____

____

____

Name ______________________________

Prediction Perfection

Scientists make predictions about things that occur in nature based on observations. Here is one example: If the sun shines, a plant will produce its own food using photosynthesis. Look around your classroom, and make some predictions of your own. The first one has been done for you.

Observation	Prediction
Example: John tips back in his chair.	**Example:** John will fall.
1. ______________________ ______________________	**1.** ______________________ ______________________
2. ______________________ ______________________	**2.** ______________________ ______________________
3. ______________________ ______________________	**3.** ______________________ ______________________
4. ______________________ ______________________	**4.** ______________________ ______________________
5. ______________________ ______________________	**5.** ______________________ ______________________
6. ______________________ ______________________	**6.** ______________________ ______________________

Name

Creative Classifying

Take all the items out of your backpack, and spread them on the desk. You will need to group and classify all these items.

1. List all the items that were in your backpack.

2. Into how many groups can you organize the items?

3. Name each group, and list the items that you put into each group.

4. Explain how you decided which item would go into each of the groups.

Name ______________________________

Four Properties of Matter

Fill in each line with the word from the box that best completes the sentence. You will use some words more than once.

mass	density	specific
volume	weight	matter

1. All matter has four general properties. These are ______________, ______________, ______________, and ______________.

2. Color, odor, size, shape, texture, and hardness are ______________ properties of matter.

3. The amount of matter in an object is the object's ______________.

4. The ______________ of an object is dependent on the amount of gravity that is acting upon it.

5. The ______________ of an object is the amount of space that the object occupies.

6. ______________ is the mass that occupies a certain volume of an object.

7. All objects are made of ______________.

Name ______________________________

Phases of Matter

Write the letter of the correct answer on the line provided.

1. A dog, a rock, and a chair are all examples of ____.
2. Definite shape and definite volume are properties of a ____.
3. Solids that have a repeating pattern are called ____.
4. A ____ has a definite volume, but no definite shape.
5. A liquid will ____.
6. A ____ will fill all available space in a container.
7. A gas has ____.

a. no definite shape or volume

b. crystals

c. take the shape of its container

d. solids

e. liquid

f. solid

g. gas

Name ______________________________

Phase Changes

Answer the following in complete sentences.

1. Compare and contrast freezing point and melting point.

____
____
____
____

2. Define condensation. Give one example. ______________________________

____
____

3. Define vaporization. ______________________________

____
____
____

4. Compare and contrast evaporation and boiling.

____
____
____
____
____

Name ____

Classes of Matter

Complete the chart below. Then answer the questions that follow.

Mixtures
Definition:
Examples:
Compounds
Definition:
Examples:
Elements
Definition:
Examples:

1. In the morning when you pour milk on your cereal, are you creating a mixture, a compound, or an element? ____________________

2. Are the substances on the periodic table elements, compounds, or mixtures? ____________________

3. Is table salt an example of a mixture, a compound, or an element?

____

Name ____________________

The Atom

Answer the following.

1. What is an atom? ____________________

____

2. What are the parts of an atom? ____________________

____

3. Electrons have a ____________ charge.

4. Neutrons have a ____________ charge.

5. Protons have a ____________ charge.

6. Which two parts of an atom are located in the nucleus? ____________

____

7. Where are the electrons located? ____________________

____

8. In the space below, draw a diagram of an atom. Label the parts of it.

Name ______________________________

Element or Molecule?

The periodic table of the elements lists every element that has been discovered on Earth. Use the periodic table below to determine if the following substances are molecules or elements. Write **M** on the line next to each molecule and **E** on the line next to each element.

Periodic Table of the Elements

Key

6 ← Atomic Number
C ← Element's Symbol
Carbon ← Name

1 H Hydrogen																	2 He Helium
3 Li Lithium	4 Be Beryllium											5 B Boron	6 C Carbon	7 N Nitrogen	8 O Oxygen	9 F Fluorine	10 Ne Neon
11 Na Sodium	12 Mg Magnesium											13 Al Aluminum	14 Si Silicon	15 P Phosphorus	16 S Sulfur	17 Cl Chlorine	18 Ar Argon
19 K Potassium	20 Ca Calcium	21 Sc Scandium	22 Ti Titanium	23 V Vanadium	24 Cr Chromium	25 Mn Manganese	26 Fe Iron	27 Co Cobalt	28 Ni Nickel	29 Cu Copper	30 Zn Zinc	31 Ga Gallium	32 Ge Germanium	33 As Arsenic	34 Se Selenium	35 Br Bromine	36 Kr Krypton
37 Rb Rubidium	38 Sr Strontium	39 Y Yttrium	40 Zr Zirconium	41 Nb Niobium	42 Mo Molybdenum	43 Tc Technetium	44 Ru Ruthenium	45 Rh Rhodium	46 Pd Palladium	47 Ag Silver	48 Cd Cadmium	49 In Indium	50 Sn Tin	51 Sb Antimony	52 Te Tellurium	53 I Iodine	54 Xe Xenon
55 Cs Cesium	56 Ba Barium	71 Lu Lutetium	72 Hf Hafnium	73 Ta Tantalum	74 W Tungsten (Wolfram)	75 Re Rhenium	76 Os Osmium	77 Ir Iridium	78 Pt Platinum	79 Au Gold	80 Hg Mercury	81 Tl Thallium	82 Pb Lead	83 Bi Bismuth	84 Po Polonium	85 At Astatine	86 Rn Radon
87 Fr Francium	88 Ra Radium	103 Lr Lawrencium	104 Rf Rutherfordium	105 Db Dubnium	106 Sg Seaborgium	107 Bh Bohrium	108 Hs Hassium	109 Mt Meitnerium									

* Lanthanide series:	57 *La Lanthanum	58 Ce Cerium	59 Pr Praseodymium	60 Nd Neodymium	61 Pm Promethium	62 Sm Samarium	63 Eu Europium	64 Gd Gadolinium	65 Tb Terbium	66 Dy Dysprosium	67 Ho Holmium	68 Er Erbium	69 Tm Thulium	70 Yb Ytterbium
† Actinide series:	89 †Ac Actinium	90 Th Thorium	91 Pa Protactinium	92 U Uranium	93 Np Neptunium	94 Pu Plutonium	95 Am Americium	96 Cm Curium	97 Bk Berkelium	98 Cf Californium	99 Es Einsteinium	100 Fm Fermium	101 Md Mendelevium	102 No Nobelium

1. table salt ____
2. gold ____
3. hydrogen ____
4. chlorine ____
5. ammonia ____
6. water ____
7. carbon dioxide ____
8. neon ____
9. baking soda ____
10. fluorine ____
11. sodium ____
12. sugar ____
13. silver ____
14. calcium ____
15. carbon ____

Name ________________________________

Get to Know Me?

Use the periodic table below to answer the following.

Periodic Table of the Elements

Key
6 ← Atomic Number
C ← Element's Symbol
Carbon ← Name

1 **H** Hydrogen																	2 **He** Helium
3 **Li** Lithium	4 **Be** Beryllium											5 **B** Boron	6 **C** Carbon	7 **N** Nitrogen	8 **O** Oxygen	9 **F** Fluorine	10 **Ne** Neon
11 **Na** Sodium	12 **Mg** Magnesium											13 **Al** Aluminum	14 **Si** Silicon	15 **P** Phosphorus	16 **S** Sulfur	17 **Cl** Chlorine	18 **Ar** Argon
19 **K** Potassium	20 **Ca** Calcium	21 **Sc** Scandium	22 **Ti** Titanium	23 **V** Vanadium	24 **Cr** Chromium	25 **Mn** Manganese	26 **Fe** Iron	27 **Co** Cobalt	28 **Ni** Nickel	29 **Cu** Copper	30 **Zn** Zinc	31 **Ga** Gallium	32 **Ge** Germanium	33 **As** Arsenic	34 **Se** Selenium	35 **Br** Bromine	36 **Kr** Krypton
37 **Rb** Rubidium	38 **Sr** Strontium	39 **Y** Yttrium	40 **Zr** Zirconium	41 **Nb** Niobium	42 **Mo** Molybdenum	43 **Tc** Technetium	44 **Ru** Ruthenium	45 **Rh** Rhodium	46 **Pd** Palladium	47 **Ag** Silver	48 **Cd** Cadmium	49 **In** Indium	50 **Sn** Tin	51 **Sb** Antimony	52 **Te** Tellurium	53 **I** Iodine	54 **Xe** Xenon
55 **Cs** Cesium	56 **Ba** Barium	71 **Lu** Lutetium	72 **Hf** Hafnium	73 **Ta** Tantalum	74 **W** Tungsten (Wolfram)	75 **Re** Rhenium	76 **Os** Osmium	77 **Ir** Iridium	78 **Pt** Platinum	79 **Au** Gold	80 **Hg** Mercury	81 **Tl** Thallium	82 **Pb** Lead	83 **Bi** Bismuth	84 **Po** Polonium	85 **At** Astatine	86 **Rn** Radon
87 **Fr** Francium	88 **Ra** Radium	103 **Lr** Lawrencium	104 **Rf** Rutherfordium	105 **Db** Dubnium	106 **Sg** Seaborgium	107 **Bh** Bohrium	108 **Hs** Hassium	109 **Mt** Meitnerium									

* Lanthanide series:	57 ***La** Lanthanum	58 **Ce** Cerium	59 **Pr** Praseodymium	60 **Nd** Neodymium	61 **Pm** Promethium	62 **Sm** Samarium	63 **Eu** Europium	64 **Gd** Gadolinium	65 **Tb** Terbium	66 **Dy** Dysprosium	67 **Ho** Holmium	68 **Er** Erbium	69 **Tm** Thulium	70 **Yb** Ytterbium
† Actinide series:	89 **†Ac** Actinium	90 **Th** Thorium	91 **Pa** Protactinium	92 **U** Uranium	93 **Np** Neptunium	94 **Pu** Plutonium	95 **Am** Americium	96 **Cm** Curium	97 **Bk** Berkelium	98 **Cf** Californium	99 **Es** Einsteinium	100 **Fm** Fermium	101 **Md** Mendelevium	102 **No** Nobelium

Write the symbol for each element.

1. chromium ____ **2.** mercury _____ **3.** boron ____

4. phosphorus ____ **5.** lead ____ **6.** selenium ____

7. argon ____ **8.** tantalum ____ **9.** lithium ____

10. arsenic ____ **11.** iron ____ **12.** radium ____

13. potassium ____ **14.** radon ____ **15.** nickel ____

Write the name of each element.

16. Sc ____________ **17.** Cu ____________ **18.** Ge ____________

19. Pt ____________ **20.** Ir ____________ **21.** Br ____________

22. Co ____________ **23.** Xe ____________ **24.** S ____________

25. He ____________ **26.** Zn ____________ **27.** Tb ____________

28. Fr ____________ **29.** Mg ____________ **30.** O ____________

Name ____

Who Am I?

Use a periodic table from your textbook to solve the riddles below.

1. I am an element that is in the first group on the periodic table. I have an atomic mass of 6.941. ________________
2. I am an element that is in period 6 on the periodic table. I have 75 protons. ________________
3. I am an element that is in group 10 on the periodic table. There is a coin named after me. ________________
4. I am an element that is in period 3 on the periodic table. I have 17 electrons. ________________
5. I am an element that is in group 6 on the periodic table. I have an atomic mass of 183.85. ________________
6. I am an element that is in group 1 and period 5 on the periodic table. My name sounds like a red gemstone. ________________
7. I am an element that is in group 2 on the periodic table. I build strong bones. ________________
8. I am an element that is in period 4 on the periodic table. I have 20 neutrons. ________________
9. I am an element that is in group 7 and period 5 on the periodic table.

   ____
10. I am an element that is in group 18 on the periodic table. I am often used to fill balloons. ________________

Name ________________________________

Protons, Neutrons, and Electrons

Use a periodic table from your textbook and the information in the box to help you complete the chart below.

electrons = atomic number
protons = electrons = atomic number
neutrons = atomic mass rounded to nearest whole number – atomic number

	Name	Symbol	Atomic Mass	Electrons	Protons	Neutrons
1.	calcium			20		
2.		Cu	63.546			
3.	tungsten					110
4.	scandium				21	
5.		K		19		
6.		Xe				77
7.	silver	Ag				
8.		I	126.905			
9.		Pb			82	
10.	mercury			80		

Name ______________________________

Speedy Scramble

Cross out all of the Xs and Zs and unscramble the remaining letters to find the hidden words from the box.

velocity	distance	motion	speed	acceleration

1. xzcxxzcazzxelxxzoxxxrzazxxxnziezzzt ______________________
2. zzxezdezzxpzzxxxxs ______________________
3. vzzxyzzxtxeozzxixxxczzzzl ______________________
4. zzzdxxszzxczxexxaxztziznzx ______________________
5. xxmxxtzzixzxznxxxozzzo ______________________

Now, define each of the words you unscrambled from above.

6. ____

____

7. ____

____

8. ____

____

9. ____

____

10. ____

____

Name ______________________________

Newton Knowledge

Newton's laws of motion are listed below. Label each as either the first, second, or third law of motion.

1. An unbalanced force on an object causes it to accelerate in the direction of the force. ____________________

2. An object at rest stays at rest until an unbalanced force acts on it, and an object in motion stays in motion until an unbalanced force acts on it.

   ____

3. For every action by a force, there is an equal and opposite reaction.

   ____

On the lines below, give an example of something that happens around you every day that would illustrate each of Newton's laws of motion.

4. Newton's First Law of Motion: ______________________________

   ____

   ____

5. Newton's Second Law of Motion: ______________________________

   ____

   ____

6. Newton's Third Law of Motion: ______________________________

   ____

   ____

Name ______________________________

Motion Commotion

Go outside, into the hallway, out to recess, or wherever there is activity in or around your school. Choose three activities that involve motion. State the law of motion you are seeing for each activity and explain why.

1. **Activity one:** ______________________________

____

____

____

____

2. **Activity two:** ______________________________

____

____

____

____

3. **Activity three:** ______________________________

____

____

____

____

Name ______________________________

Are We There Yet?

Use the formula in the box to help solve the word problems that follow.

$$\text{average speed} = \frac{\text{total distance}}{\text{total time}}$$

1. Your family is taking a trip to the beach. The beach is 130 miles away. You need to be there in 3 hours. How fast does your mother need to drive?

2. You live 20 miles from school. You are on the school bus for 1 hour. What is the average speed of your bus?

3. Your best friend lives 10 miles from you. It takes you 2 hours to ride your bike to her house. How fast are you riding?

Name ______________________________

Amazing Acceleration

Use the formula in the box to help solve the word problems that follow.

$$\text{acceleration} = \frac{\text{final velocity} - \text{initial velocity}}{\text{time}}$$

1. Your teenage brother has a sports car. He tells his friends that his car can go from 0 mph to 60 mph in 10 seconds. What is the acceleration?

2. Your mom's minivan can go from 0 mph to 40 mph in 10 seconds. What is the acceleration?

3. When you ride your bike, you can go from 0 mph to 5 mph in 8 seconds. What is the acceleration?

Name ______________________________

Finding the Force

Use the formula in the box to help solve the word problems that follow.

force = mass × acceleration

1. At what force would you have to push your 20 kg bicycle to get it to accelerate away from you at 2 m/s^2?

2. At what force would you have to push an 8 kg rock to get it to accelerate down a hill at 3 m/s^2?

3. At what force would you have to throw a 0.10 kg baseball to get it to accelerate at 20 m/s^2?

Name ____

Force Facts

Fill in each line with the word from the box that best completes the sentence.

gravitational force	electrical force	magnetic force
friction	nuclear force	weak interactions

1. Forces that may cause the nuclei of an atom to break apart are ____________________.

2. The force that opposes motion is ____________________.

3. ____________________ is the force of attraction between two objects.

4. The force caused by moving electrical charges is ____________________.

5. ____________________ is the force that holds the nucleus of an atom together.

6. The force between electrical charges is ____________________.

Name ____

Forces of Liquids

Choose the correct term to complete each statement.

1. An object in a fluid is supported by a force equal to the weight of the fluid that is displaced. This is _____.
   - **a.** Bernoulli's Principle
   - **b.** Pascal's Law
   - **c.** Archimedes' Principle

2. The pressure of a fluid is the same at all points in a closed container. This is called _____.
   - **a.** Archimedes' Principle
   - **b.** Pascal's Law
   - **c.** Bernoulli's Principle

3. As the speed of a fluid increases, the pressure in a moving fluid decreases. This is called _____.
   - **a.** Bernoulli's Principle
   - **b.** Pascal's Law
   - **c.** Archimedes' Principle

Give an example of each principle or law listed below.

4. Archimedes' Principle: ____

   ____

5. Pascal's Law: ____

   ____

6. Bernoulli's Principle: ____

   ____

Name ______________________________

Work and Force

Using the terms in the box, unscramble each word below and define the term.

work	effort force	resistance force

1. ortffe orcef ____________ Define: ____________

____

____

2. ecretnaiss eforc ____________ Define: ____________

____

____

3. orwk ____________ Define: ____________

____

____

Use the formula in the box to answer the following questions.

work = force × distance

4. A boy uses a force of 300 N to move a rock 10 m. How much work is done?

5. A girl uses a force of 500 N to move her bed 5 m. How much work is done?

Name ____________________

Waves and Sound

Match each term to its definition. Write the letter of the correct answer on the line provided.

1. _____ acoustics
2. _____ wavelength
3. _____ reflection
4. _____ refraction
5. _____ Doppler effect
6. _____ decibel
7. _____ frequency
8. _____ intensity
9. _____ pitch
10. _____ amplitude

a. the greatest movement of matter from its normal resting position

b. As a wave passes from one medium to another, it changes speed and seems to bend.

c. the unit used to measure sound intensity

d. In sound, the change of pitch as the source of a sound gets closer to or farther away from a listener.

e. how high or low a sound is

f. the study of sound

g. the bouncing back of waves that strike a surface

h. the loudness of sound

i. the distance from any point on a wave to the same point on the next wave

j. the number of waves that pass a given point each second

Name ______________________________

Sound Waves

A **sound wave** is a kind of wave that moves energy from one place to another. The energy is transmitted through the particles of matter the sound is traveling through.

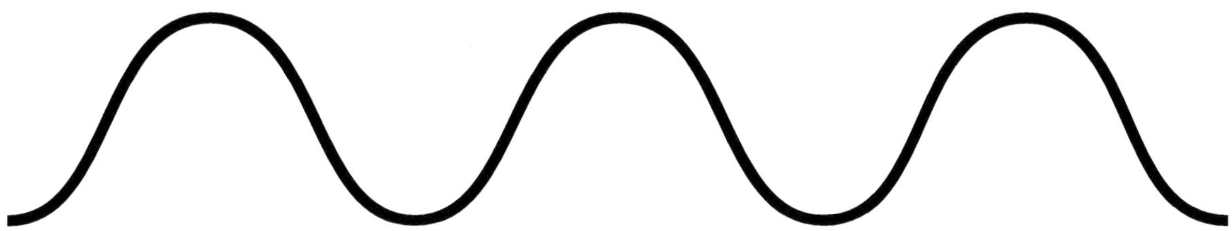

Label the following parts of the wave and define each term.

1. amplitude: ______________________________

____

2. wavelength: ______________________________

____

3. crest: ______________________________

____

4. trough: ______________________________

____

Name ____________________

Reflection and Refraction

Use the words from the box to solve the puzzles.

reflection	refraction	incident ray
index of refraction	the law of refraction	reflected ray
angle of reflection	angle of refraction	

1. _ _ _ L _   _ _   _ _ _ L _ _ _ _ _ _ _ _
2. I _ _ _ _ _ _ _   _ _ _
3. _ _ G _ _   _ _   _ _ _ R _ _ _ _ _ _ _ _
4. _ H _   _ _ _   _ _   R _ _ _ _ _ _ _ _ _
5. _ _ _ _ A _ _ _ _ _
6. R _ _ L _ _ _ _ _ _
7. I _ _ _ _   _ _   _ E _ _ _ _ _ _ _ _
8. _ _ _ _ _ _ _ _ D   _ _ Y

Define three of the terms you used above.

9. ____________________
10. ____________________
11. ____________________

Name ____

Light Logic

Fill in each line with the word from the box that best completes the sentence.

refraction	light year	convex
diffraction	light rays	incandescence
reflection	concave	

1. The bouncing back of light rays from a surface is ____________________.
2. A lens that is thinner in the center than at the edges is ____________________.
3. The distance that light travels in one year is a ____________________.
4. Beams of light are made of several ____________________.
5. The spreading out of waves past the edge of a barrier or through a small opening is called ____________________.
6. The emission of light by a heated object is called ____________________.
7. The change in direction of a light ray as it passes from one medium to another is ____________________.
8. A lens that is thicker in the middle than at the edges is ____________________.

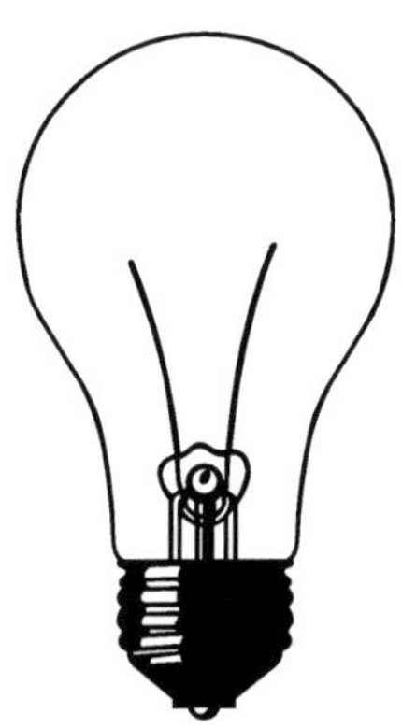

Name ______________________________

Looking at Light

Use the space provided to answer the following questions.

1. What does the law of reflection state? ______________________________

2. How are various colors of light produced? ______________________________

3. How would you compare and contrast reflection and refraction? ______

4. What is an example of an object that is opaque, an object that is transparent, and an object that is translucent? ______________________________

Name ________________________________

Let's Heat Things Up!

Match each term to its definition. Write the letter of the correct answer on the line provided.

1. ____ conduction
2. ____ temperature
3. ____ convection
4. ____ heat
5. ____ radiation
6. ____ heat of fusion
7. ____ pressure
8. ____ heat of vaporization
9. ____ kinetic theory
10. ____ specific heat

a. the measure of the average kinetic energy of the particles in a sample of matter

b. the amount of heat energy needed to raise the temperature of 1.0 grams of a substance by 1 degree Celsius

c. the transfer of energy by waves

d. a force acting on a unit area

e. a theory stating that all matter is made of particles that are in constant motion

f. the amount of heat energy that will change a liquid at its boiling point to a gas at the same temperature

g. the amount of heat energy that will change a solid at its melting point to a liquid at the same temperature

h. the total energy of all the particles in a sample of matter

i. the heat transfer that takes place in fluids

j. the transfer of heat by direct contact between particles

Name ____

Hot and Cold

After each statement, write *addition of heat* or *loss of heat*.

1. A pot of pasta is being cooked for dinner. ______________________
2. A tray of water freezes into ice cubes. ______________________
3. The ice on a lake is melting in the spring. ______________________
4. Clothes are being dried in a clothes dryer. ______________________
5. A frozen steak is thawed in a microwave. ______________________

In the space provided, give an example of loss of heat and an example of the addition of heat.

6. Example of loss of heat: ______________________

   ____

   ____

7. Example of the addition of heat: ______________________

   ____

   ____

Name ______________________________

Gas Laws

Throughout history, scientists have been interested in the properties that control how hot-air and gas-filled balloons work. Scientists, such as Jacques Charles and Robert Boyle, investigated the relationships between pressure, volume, and temperature.

Label each gas law as either Charles' Law or Boyle's Law.

1. The volume of a gas decreases as its pressure increases if the temperature stays the same. The volume of a gas increases as its pressure decreases if the temperature stays the same.

   ____

2. The volume of a gas increases as its temperature increases if the pressure stays the same. If the temperature of a gas decreases while the pressure stays the same, the volume of the gas decreases.

   ____

Think about the world around you, and give one real-life example of Boyle's Law and one real-life example of Charles' Law.

3. Boyle's Law ______________________________

   ____

   ____

4. Charles' Law ______________________________

   ____

   ____

Name ______________________________

Energy for Engines

Answer the following in the space provided.

1. In an external combustion engine, the fuel is burned ____________ the engine.
2. Give one example of an external combustion engine. ____________
____
3. In an internal combustion engine, the fuel is burned ____________ the engine.
4. Give one example of an internal combustion engine. ____________
____
5. Explain how each step in a four-stroke gasoline engine works.
   a. Intake stroke: ______________________________
   ____
   ____
   b. Compression stroke: ______________________________
   ____
   ____
   c. Power stroke: ______________________________
   ____
   ____
   d. Exhaust stroke: ______________________________
   ____
   ____
6. What is the major difference between a four-stroke gasoline engine and a diesel engine? ______________________________
____
____

Name ____

Magnet Magic

Use the words in the box to help you unscramble the terms. Then write a definition for each term.

electromagnet	magnet	magnetic field
lines of force	poles	

1. tmeagn ____________— ______________________________________

____

____

2. esolp _________— ____

____

____

3. eisln fo ceorf _________ ____ _________— __________________

____

____

4. neticgam ldief _______________ _________— __________________

____

____

5. getnamtroleec _____________________— ______________________

____

____

Name ______________________________

Scientists in the News

Imagine that you live during the time of each scientist below. Write a short newspaper article that describes the work of each of the scientists. Use your textbook or research materials, if necessary.

1. Hans Christian Ørsted: ______________________________

____

____

____

____

____

2. Michael Faraday: ______________________________

____

____

____

____

____

Name ______________________________

Magnetic Mania

1. Obtain a magnet from your teacher. In the space below, record all of the items in your classroom that are magnetic in one column and all the items you found that are not magnetic in the other column.

Magnetic	Not Magnetic
______________________	______________________
______________________	______________________
______________________	______________________
______________________	______________________
______________________	______________________
______________________	______________________
______________________	______________________
______________________	______________________
______________________	______________________
______________________	______________________
______________________	______________________
______________________	______________________
______________________	______________________

2. What do all of the things that are magnetic have in common? ________

____

____

Name ____________________

Electricity Puzzler

Fill in each line with the word from the box that best completes the sentence.

anode	electric circuit	kilowatt
cathode	induction	voltage
conductor	insulator	watt
		resistance

1. ____________________ is the opposition to the flow of charges in a substance.
2. The electrode through which electrons leave an electric cell is the ____________________.
3. A(n) ____________________ does not allow electric charges to flow through easily.
4. The ____________________ is the SI unit of power.
5. The electrode through which electrons enter an electric cell is the ____________________.
6. ____________________ is the measure of the energy available to move charges in a circuit.
7. A(n) ____________________ is a substance that allows electric current to flow through easily.
8. A(n) ____________________ is a unit of power equal to 1,000 watts.
9. ____________________ is the process of charging an object by the presence of a nearby charged object.
10. A system through which an electric current can travel is an ____________________.

Name ____

All About Electricity

In the space provided, write the letter of the answer that correctly completes each statement.

1. There are two types of electrical charges. They are _____.
   - a. positive and neutral
   - b. positive and negative
   - c. negative and neutral

2. Two negatively charged items will _____ each other when brought close together.
   - a. attract
   - b. melt
   - c. repel

3. An English scientist named _____ discovered that metal wire could conduct an electric charge.
   - a. Stephen Gray
   - b. Thales
   - c. Thomas Edison

4. When an object gains or loses electrons, _____ electricity is produced.
   - a. current
   - b. static
   - c. alternating

5. When electric charges move through a wire, _____ electricity is produced.
   - a. static
   - b. current
   - c. alternating

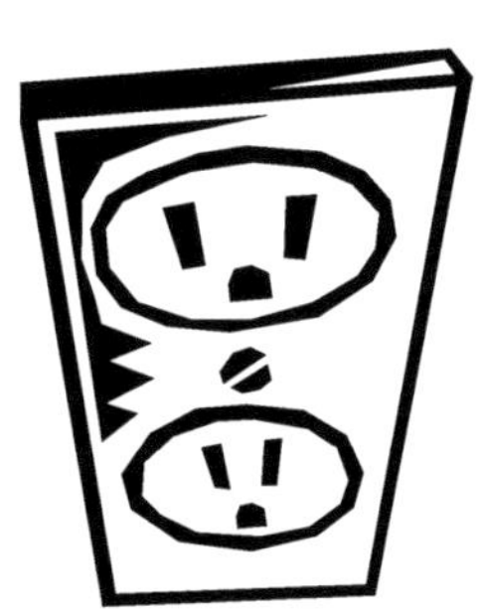

Name ______________________________

Circuit Secrets

Explain how each of the circuits listed below conducts electricity.

1. Parallel circuit: ______________________________

____

____

2. Series circuit: ______________________________

____

____

3. Printed circuit: ______________________________

____

____

Give an example of how each type of circuit is used.

4. Parallel circuit: ______________________________

____

5. Series circuit: ______________________________

____

6. Printed circuit: ______________________________

____

Name ____

Super Scientists

Write a brief summary describing the contributions that each of the following scientists made to the study of electricity. Use your textbook or research materials, if necessary.

1. Charles Proteus Steinmetz: ____

____

____

____

____

2. Stephen Gray: ____

____

____

____

____

3. Georg S. Ohm: ____

____

____

____

____

Name ________________________________

Common Sense

All living things have three common characteristics that set them apart from nonliving things. They need food, they grow and change, and they come from the same kind of living thing.

Compare and contrast these pairs of living things.

1. butterfly and snake: ______________________________

2. dog and cat: ______________________________

3. eagle and chipmunk: ______________________________

Name ______________________________

Similar Structures

All living things have structures that are used to classify them into different groups. For example, a spruce tree and a maple tree have similar structures. These may include bark, leaves, and roots. However, they also have structures that differ. For example, maple trees have broad flat leaves, and spruce trees have narrow needles.

List the similar structures for the pairs of animals below. Also, list some of the structures that are different for each pair.

1. eagle and robin: ______________________________

____

____

____

____

2. chipmunk and squirrel: ______________________________

____

____

____

____

3. trout and shark: ______________________________

____

____

____

____

Name ____

Cell Game

Match the person or term to its definition or description. Write the letter of the correct answer on the line provided.

1. ____ Anton van Leeuwenhoek

2. ____ organelles

3. ____ cells

4. ____ Robert Hooke

5. ____ Mathias Schleiden

6. ____ cell theory

7. ____ Theodor Schwann

8. ____ Rudolf Virchow

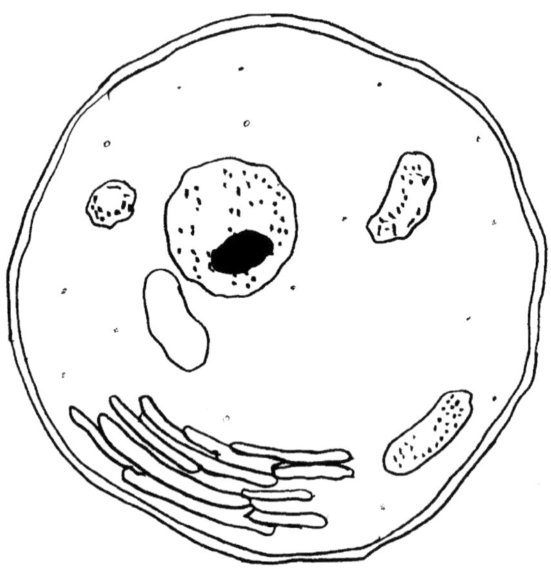

a. the basic unit of structure of all living things

b. discovered that all living cells come only from other living cells

c. All living things are made of cells; cells are the basic unit of structure and function of living things; living cells come only from other living cells.

d. proposed that all plants are made of cells

e. the first scientist to use the term *cell*

f. proposed that all animals are made of cells

g. found tiny living things in a drop of rainwater

h. the tiny structures that make up a cell

Name ______________________________

Calling All Cells!

Fill in each line with the correct term from the box below.

chloroplast	protoplasm	endoplasmic reticulum
cell membrane	nucleus	nucleolus
ribosomes	lysosome	nuclear membrane
mitochondria	chromosomes	cell wall
vacuoles		

1. The outermost layer of a plant cell. ____________________
2. A thin, flexible membrane that surrounds all cells.

   ____
3. All living material in a cell. ____________________
4. The control center of the cell. ____________________
5. The membrane that separates the nucleus from the rest of the cell.

   ____
6. The material that is in the nucleus of the cell and contains the genetic material. ____________________
7. The organelle that contains the RNA. ____________________
8. The organelle that is involved in the transportation of proteins.

   ____
9. The protein-making sites in the cell. ____________________
10. The organelle that supplies the energy to the cell.

    ____
11. The storage tanks of the cell. ____________________
12. The structure that is involved in the digestion of materials in the cell.

    ____
13. The organelle that contains the chlorophyll in plant cells.

    ____

Name ______________________________

Animal Cell

Label the animal cell below using the terms from the box.

lysosome	nucleus	mitochondria
ribosome	vacuole	nuclear membrane
cell membrane	endoplasmic reticulum	nucleolus

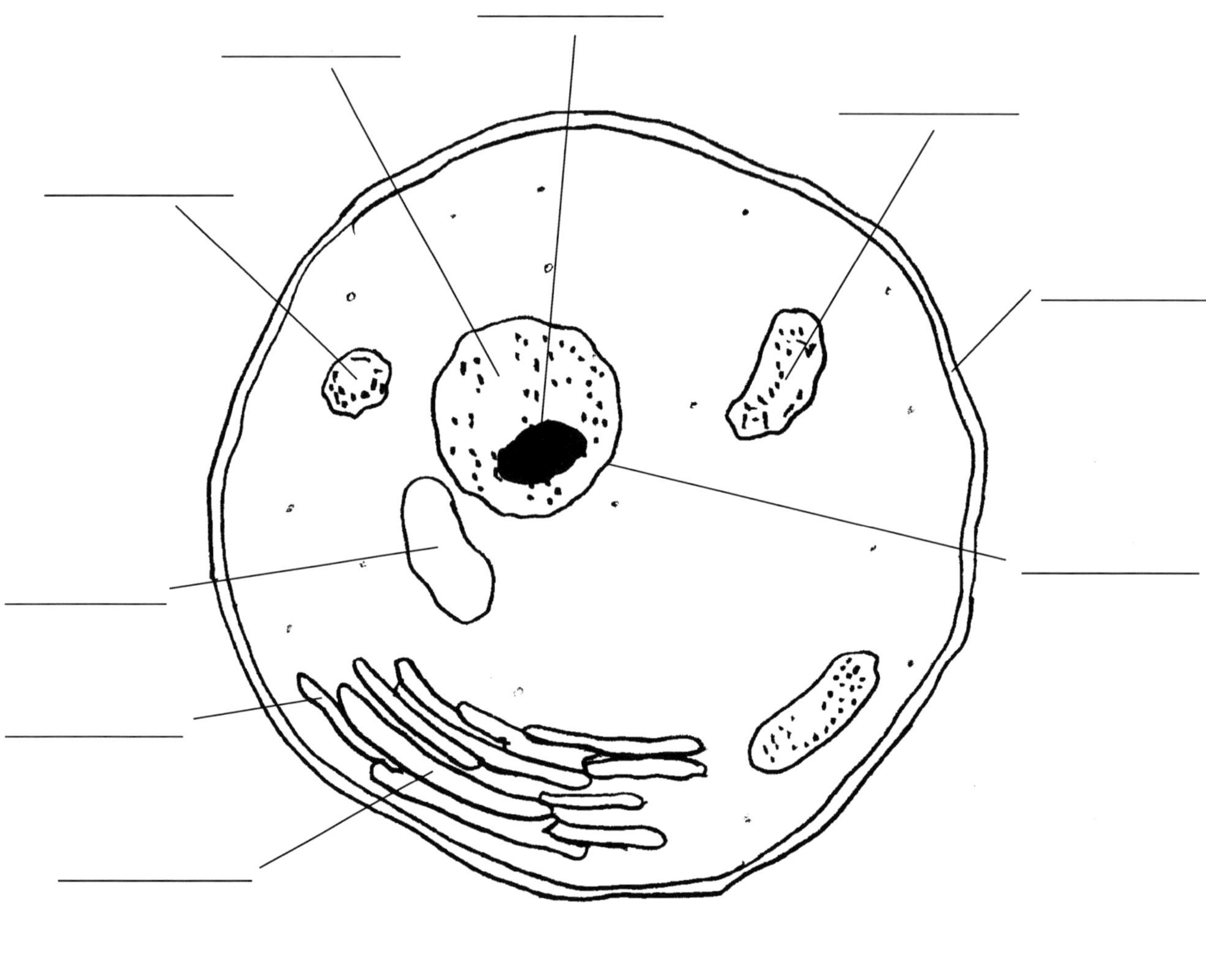

Name ______________________________

Plant Cell

Label the diagram of the plant cell below using the terms from the box.

vacuole	cell membrane	ribosomes
mitochondria	cell wall	nucleolus
nucleus	endoplasmic reticulum	nuclear membrane
chloroplasts		

Name ______________________________

Plant Versus Animal

Imagine that a plant cell and an animal cell are arguing about which of them is more complicated. In the space below write what they would say. Make reference to the organelles and the functions that they serve.

Name ____________________

Get Moving!

Choose a living thing (such as a cat, a dog, a tree, a bird, or any other living thing), and write a description of how it would do each of the following life activities.

living thing you chose: ____________________

getting food: ____________________

____

____

movement: ____________________

____

____

growing: ____________________

____

____

sensing and reacting: ____________________

____

____

using food and removing wastes: ____________________

____

____

development: ____________________

____

____

reproduction: ____________________

____

____

Name ______________________________

King Me

Read through the list of organisms in the box below. Write the name of each organism under its correct kingdom.

dog	yeast	daisy	E. coli
puffball	eel	salmonella	sporozoan
frog	ant	penicillin	grass
rose	amoeba	oak tree	streptococcus
mushroom	spirillum	euglena	paramecium

Plant

____

____

____

____

Animal

____

____

____

____

Protist

____

____

____

____

Fungi

____

____

____

____

Bacteria

____

____

____

____

Name ______________________________

Animal Kingdom

The animal kingdom contains both animals with backbones, called **vertebrates,** and animals without backbones, called **invertebrates.** List five animals that are in each group.

Invertebrates:

1. ______________
2. ______________
3. ______________
4. ______________
5. ______________

Vertebrates:

1. ______________
2. ______________
3. ______________
4. ______________
5. ______________

In the box below, there are seven levels that scientists use to classify each animal. Place these seven levels in order from largest level to smallest level.

species	family
kingdom	class
genus	order
phylum	

1. ______________
2. ______________
3. ______________
4. ______________
5. ______________
6. ______________
7. ______________

Name ______________________________

Plant Kingdom

Fill in each line with the word from the box that best completes the sentence.

conifers	vascular	dicots
nonvascular	cone	parallel
monocot	network	gymnosperms
mosses		

1. ________________ plants must be in contact with water at all times because they cannot transport water.
2. Plants that can transport water are called ________________.
3. Beans are ________________ because their seeds have two halves.
4. The veins in the leaves of monocots are ________________.
5. The seed of a ________________ is in one piece.
6. The leaves of dicots have a ________________ of veins.
7. Plants that do not flower are called ________________.
8. The seeds of a gymnosperm are produced inside of a ________________.
9. The major group of gymnosperms are the ________________.
10. ________________ are nonvascular plants.

Name ____________________

Protist Puzzler

Answer these questions in complete sentences.

1. What is the major difference between animallike and plantlike protists?

____
____
____

2. What are three features of plantlike protists?

____
____
____

3. Most plantlike protists are algae. Can you name four groups of algae?

____
____
____

4. How does an amoeba move?

____
____
____

5. What is an example of a funguslike protist?

____
____
____

Name ______________________________

Fungus Features

Use a textbook or research materials, if necessary, to answer the following questions about the fungus of your choice. Circle the fungus that you chose. Answer the questions in complete sentences.

mold	penicillin	mushrooms	yeast

1. Why does the fungus you chose classify as a fungus?

____

____

____

____

2. Where does the fungus grow?

____

____

____

____

3. What are some uses of the fungus?

____

____

____

____

____

____

Name ______________________________

Behavior Basics

Circle the correct response to each of the following.

1. Innate behavior is a behavior that an animal learns.
   a. true **b.** false

2. Nest building is an innate behavior for what type of animal?
   a. cats **b.** frogs **c.** birds

3. What is the purpose of courtship behaviors?
   a. to get food **b.** to get a mate **c.** to get shelter

4. The purpose of territorial behavior is to ensure a good food supply, find a mate, and reduce fighting.
   a. true **b.** false

5. Imprinting behavior is instant learning that is common among chicks and goslings.
   a. true **b.** false

6. What is phototropism?
   a. a plant behavior to get light
   b. behavior of bees
   c. animal-mating behavior

Name ______________________________

Learned Behavior

A **learned behavior** is a behavior that involves the use of thinking and reasoning. These behaviors include responses such as fish moving to the surface of the water at feeding time or cows going to the barn at milking time. Think of an example of a learned behavior that you have witnessed in an animal. Use a textbook or research materials to find more examples of learned behavior, if needed. Explain your example, and discuss the training that probably took place for the animal to learn this behavior.

Name ____________________

Communication

Communication is a valuable animal behavior that enables animals to trade information with one another. This behavior aids in the survival of all animals. Think about how human survival would be in danger if humans could no longer communicate. No communication means no reading, writing, talking, or physical gestures. Write a paragraph describing what impact you believe a lack of communication would have on human existence.

Name ______________________________

Food Fun

Animals obtain food in a variety of ways. Match each animal with the way in which it obtains food. Write the letter of the correct answer on the line provided.

1. ____ Japanese beetle
2. ____ toad
3. ____ housefly
4. ____ butterfly
5. ____ hydra
6. ____ sponge
7. ____ wolf
8. ____ spiders

a. hunts prey in packs

b. has chewing parts for eating leaves

c. pumps water into the body and filters it for microorganisms

d. catches prey in webs

e. catches prey with a long, sticky tongue

f. has sucking mouthparts to siphon nectar

g. has stinging tentacles to sting and capture prey

h. has piercing mouthparts to suck blood

Name ______________________________

Digestion Question

Fill in each line with the word from the box that best completes the sentence.

fangs	small intestine	gall bladder	gizzard	stomach

1. Spiders use a pair of ____________________ to inject digestive enzymes into their prey.
2. The ____________________ is an organ found in insects and birds in which food is ground up.
3. The ____________________ stores and releases bile that breaks down fats.
4. After food is swallowed, it enters a sack of digestive enzymes called the ____________________.
5. When food enters the ____________________, bile from the liver and enzymes from the pancreas continue to break down food.

6. Use a textbook or research materials to help describe the path that food takes when eaten by a person. On the lines below, explain the process step by step.

____

____

____

____

____

____

____

Name ______________________________

Respiration and Circulation

The circulation and respiration of the three different animals that are listed in the box is described below. On the line provided, write the name of the animal that is being described.

jellyfish	insect	frog

1. This animal has a three-chambered heart. Within this animal, there are two kinds of circulation. One kind is circulation to and from the lungs, and the other is circulation to and from the body. The animal's cells use oxygen to carry out cellular respiration. The animal spends a lot of time in water so respiration can also occur through the animal's thin skin.

   ____

2. The cells of this animal are in direct contact with the water. Circulation of the water carries materials in and out of the body. Waste materials from respiration are diffused from the cells to the body cavity. The wastes are then expelled through the mouth.

   ____

3. The circulatory system in this animal is an open system. There is a heart that is a swelling of a tube called the aorta. Small tubes run along the sides of the animal and open and close to a rhythm. These tubes open directly to the outside air.

   ____

Name ______________________________

Nervous System Scramble

Under the name of each animal, list the parts of its nervous system. Use the terms in the box to help you. (Some of the terms will be used more than once.)

brain	compound eye	chain of autonomic ganglia
antennae	medulla oblongata	ventral nerve cord
cerebrum	brachial nerve	cranial nerves
nerves	optic lobe	spinal nerves
ganglia	sciatic nerve	spinal cord
cerebellum	brain nerves	

Frog

____
____
____
____
____
____
____
____
____
____
____

Earthworm

____
____
____
____

Grasshopper

____
____
____
____
____
____

Name ____________________

Plant Parts

Answer the questions using complete sentences.

1. How does the root of a plant work?

____

____

____

____

2. What are the parts of the root called?

____

____

____

3. What is the job of the stem?

____

____

____

4. What are the parts of a leaf called?

____

____

____

Name ____

The Story of Photosynthesis

Read the story of photosynthesis below. Fill in each line with the word from the box that best completes the sentence.

chlorophyll	oxygen	roots	carbon dioxide
chloroplast	hydrogen	oxygen	sugars

Sunlight hits the leaf of a plant. The ________________ in the leaf captures the sunlight and keeps it. The chlorophyll is contained in a(n) ________________. Water enters the plant through the ________________. As the sun hits the leaf, it helps the plant split the water into two parts: ________________ and ________________. The hydrogen then combines with ________________ to form simple sugars. The plant uses the ________________ as food. As the process of photosynthesis takes place, the plant releases ________________ into the air. The oxygen is then used by animals.

Name ____

The Oxygen-Carbon Dioxide Cycle

In the space below, describe the oxygen-carbon dioxide cycle. Use complete sentences. Then draw a diagram that illustrates the cycle.

Description:

____

____

____

____

____

____

Diagram:

Name ____

Plant Reproduction

Write **T** for **true** or **F** for **false** after each statement.

1. One-celled organisms divide into two cells, and each cell becomes a new organism. This is known as asexual reproduction. _____

2. During sexual reproduction, an egg cell and a sperm cell join to form a zygote. _____

3. Mosses and ferns reproduce only sexually. _____

4. Seed plants reproduce asexually. _____

5. The egg cells in seed plants are encased in a structure called the ovary. _____

6. The flower of an angiosperm produces both the egg cells and the sperm cells. _____

7. In a flowering plant, the stamen is the female reproductive organ and the pistil is the male reproductive organ. _____

8. Seeds contain the embryo of a new plant. _____

9. Water is not the key factor in the beginning of seed germination. _____

10. Cones contain the reproductive organs in gymnosperms. _____

Name ____________________

Life Cycles

Place the stages of development for a frog and a butterfly in order. Use a textbook or an encyclopedia to help you.

Frog

egg

tadpole

embryo

legs develop

1. ____________________
2. ____________________
3. ____________________
4. ____________________
5. adult frog

Butterfly

pupa

egg

larva

1. ____________________
2. ____________________
3. ____________________
4. adult butterfly

Name ________________________________

A Chip off the Old Block

Answer the following in complete sentences.

1. What is DNA, and why do we need it?

____
____
____
____

2. Think about how animal populations can be different. Explain how these differences can help animal populations.

____
____
____
____

3. Explain why animals reproduce, and describe what would happen if animals did not have offspring.

____
____
____
____

Name ____________________

Heredity

Match each term to its definition. Write the letter of the correct answer on the line provided.

1. _____ genetics
2. _____ trait
3. _____ self-pollination
4. _____ cross-pollination
5. _____ dominant genes
6. _____ recessive genes
7. _____ hybrid
8. _____ incomplete dominance
9. _____ phenotype
10. _____ genotype

a. the genetic makeup of an organism

b. the characteristic of an organism

c. the condition that occurs when a gene is neither dominant nor recessive

d. the process in which pollen is transferred from the male organ of one plant to the female organ of another plant

e. the genes of a stronger trait

f. an organism produced through hybridization

g. a visible characteristic of an organism

h. the process in which a plant pollinates itself

i. the genes of a weaker trait

j. the study of heredity

Name ____

Mendel's Peas

Gregor Mendel is called the Father of Genetics. Use a textbook or reference materials to help answer the following about him.

1. What is the hypothesis that Mendel formed about how traits are inherited?

____

____

____

2. What happened when Mendel crossed two true-breeding tall pea plants?

____

____

____

3. What happened when Mendel crossed a pure tall pea plant with a pure short pea plant?

____

____

____

4. Explain how his findings proved his hypothesis.

____

____

____

Name ____________________

Chromosomes

Answer the questions below in complete sentences.

1. What is a mutation?

____

____

____

____

2. Can you give one example of a genetic mutation caused by the mutation of the chromosomes? (Example: Down's syndrome is a genetic mutation caused by the alteration of a chromosome. Use reference materials, if needed.)

____

____

____

____

3. What is the main function of chromosomes?

____

____

____

____

Name ____________________

Biotic and Abiotic

Fill in each line with the word from the box that best completes the sentence.

biotic	ecology	abiotic	animals

1. An environment is a web of intertwining threads. The threads of an environment consist of plants, ____________________, soil, water, temperature, and light.
2. The living things in an environment are ____________________ factors.
3. The nonliving things in an environment are ____________________ factors.
4. ____________________ is the study of the relationship and interactions of living things.
5. List all of the abiotic factors and all of the biotic factors that exist in your classroom environment.

Biotic Factors	Abiotic Factors

Name

Ecosystem Essay

Think of an ecosystem that you have visited. It could be a field, a forest, a pond, the ocean, and so forth. In the space below, write a descriptive essay about the ecosystem and all of the living and nonliving things in the environment. Tell how they depend on one another.

Name ______________________________

Ecosystem Equation

There are three parts to an ecosystem: the community, the population, and the habitat. Choose an animal and describe the community, the population, and the habitat of the animal. (Example: A white-tailed deer lives in a forest. That is the community. Its population includes all of the white-tailed deer living in the forest. A sheltered area for sleeping and a grassy area for feeding make up the habitat of a white-tailed deer.)

community + population + habitat = ecosystem

animal you chose: ______________________________

community: ______________________________

____

____

population: ______________________________

____

____

habitat: ______________________________

____

____

Name ______________________________

Habitat Match

A habitat is a place in which an organism lives. Match each animal to its habitat. Write the letter of the correct habitat on the line provided.

1.	____ river trout	a.	arctic tundra
2.	____ flounder	b.	grassy field
3.	____ bullfrog	c.	pond
4.	____ woodchuck	d.	river
5.	____ white-tailed deer	e.	ocean
6.	____ jaguar	f.	rain forest
7.	____ lion	g.	African grasslands
8.	____ polar bear	h.	deciduous forest

Name ______________________________

What Do I Eat?

Organisms can be classified into three main groups based on how they obtain food that is used for energy. Define each of the terms below, and give an example of an animal for each.

1. producer: ______________________________
____
____

2. consumer: ______________________________
____
____

3. herbivore: ______________________________
____
____

4. omnivore: ______________________________
____
____

5. carnivore: ______________________________
____
____

6. decomposer: ______________________________
____
____

Name

Food Chain Fun

A food chain shows how the organisms that live together in a certain environment rely on one another for survival. (For example: In a field, an insect would feed on the grass; a bird would feed on the insect; a fox would feed on the bird; the waste from the animals would nourish the soil that grows the grass.)

Draw a simple food chain for each of the habitats below.

1. freshwater lake:

2. deciduous forest:

3. ocean:

Name ______________________________

Relationship Report

Use a textbook or reference materials to find an example of each type of relationship that exists between organisms in ecosystems. (Example: A cat hunting a mouse is an example of predation.) Describe each of the relationships below with your example.

1. competition: ______________________________

____

____

2. predation: ______________________________

____

____

3. commensalism: ______________________________

____

____

4. mutualist: ______________________________

____

____

5. parasitism: ______________________________

____

____

Name

Cycle Circles

In the space provided below, draw a simple diagram that illustrates each of these cycles found in nature.

Nitrogen Cycle

Water Cycle

Oxygen-Carbon Dioxide Cycle

Name ____

Biome Buster

Scientists have classified areas with similar climates, plants, and animals as **biomes.** For example, you would find similar types of plants and animals in any desert in the world. In the space provided, give a brief description of each of the eight biomes.

1. tundra: ______________________________

____

2. coniferous forest: ______________________________

____

3. deciduous forest: ______________________________

____

4. tropical rain forest: ______________________________

____

5. grasslands: ______________________________

____

6. desert: ______________________________

____

7. ocean: ______________________________

____

8. freshwater: ______________________________

____

Name ____

Biome Matchup

Match each term to its definition. Write the correct letter on the line provided.

1. _____ biome
2. _____ canopy
3. _____ climate
4. _____ taiga
5. _____ savanna
6. _____ permafrost
7. _____ phytoplankton

a. a roof formed by tall trees in a forest

b. the northernmost area of a coniferous forest biome

c. grassland

d. the average weather in a particular place over a long period of time

e. the division of area with similar climate, plants, and animals

f. the microscopic plants that live at the surface of the ocean

g. permanently frozen tundra soil

Name ____

My Favorite Biome

Choose one of the eight biomes from the box, and use a textbook or reference materials to give a detailed description of it. Include a description of the climate, the plants, and the animals.

tundra	deciduous forest	grasslands	ocean
coniferous forest	tropical rain forest	desert	freshwater

Biome you chose: ______________________________

____

____

____

____

____

____

____

____

____

____

Draw a picture that illustrates the biome you chose.

Name ____________________

Pollution

Listed below are three types of pollution that are caused by humans. In the space provided, give as many examples as possible of how humans cause the pollution.

1. air pollution: ____________________

____

____

____

2. water pollution: ____________________

____

____

____

3. noise pollution: ____________________

____

____

____

Name ______________________________

What Am I Doing?

There are many state and national parks all over the country. The land that is part of these parks was once untouched by humans. The humans that visit these parks each year have a great amount of impact on these ecosystems. Imagine that you are an animal living in a state or national park. Write a brief story describing what the human visitors are doing to your home.

Name ______________________________

Earth Materials

In the space provided, answer the questions using complete sentences.

1. *Lithosphere* is the word used to describe what part of Earth?

____

____

2. What does the Greek word *litho* mean?

____

____

3. What is Earth's hydrosphere?

____

____

4. What does the Greek word *hydro* mean?

____

____

5. What is Earth's atmosphere?

____

____

6. What does the Latin word *atmos* mean?

____

____

Name

My Earth

In the space provided, draw a cross section diagram of Earth, and label its sections using the terms in the box.

lithosphere	asthenosphere	mantle
atmosphere		core

Name ____

Crystal Creations

In the space provided, list the six shapes of crystals. Describe each one, and give an example.

1. ____________ : ____________________________

____

____

2. ____________ : ____________________________

____

____

3. ____________ : ____________________________

____

____

4. ____________ : ____________________________

____

____

5. ____________ : ____________________________

____

____

6. ____________ : ____________________________

____

____

Name ____________________

Mineral Matchup

Match each mineral to its definition. Write the letter of the correct definition on the line provided.

1. ____ silicate
2. ____ nonsilicate
3. ____ carbonate
4. ____ sulfate
5. ____ halide
6. ____ oxide
7. ____ sulfide
8. ____ sialic silicate
9. ____ simatic silicate
10. ____ ore mineral

a. a compound that contains chlorine or fluorine together with sodium, potassium, or calcium

b. a mineral containing units of oxygen and silicon atoms

c. a mineral that has enough of a particular element in it to be worth mining

d. a compound made up of oxygen and one or more other elements

e. a mineral that does not contain both silicon and oxygen

f. a silicate rich in magnesium and silicon

g. a compound containing a unit consisting of one carbon atom surrounded by three oxygen atoms

h. a silicate mineral rich in silicon and aluminum

i. a compound containing one or more metals combined with sulfur

j. a compound containing a unit that is made up of one sulfur atom surrounded by four oxygen atoms

Name ______________________________

Mineral Properties

Define each property that is used to identify minerals.

1. hardness: ______________________________

2. streak: ______________________________

3. luster: ______________________________

4. cleavage: ______________________________

5. fracture: ______________________________

6. specific gravity: ______________________________

Name ______________________________

Rocks On!

Answer the following using complete sentences.

1. Name the three major rock groups, and briefly describe how each is formed.

____

____

____

____

2. Name two types of igneous rocks, and compare and contrast them.

____

____

____

____

3. Describe the three ways that sedimentary rocks are formed, and name one rock formed by each of the three kinds of sediment.

____

____

____

____

4. Name the two forces that produce metamorphic rock, and explain how each occurs.

____

____

____

____

Name ______________________________

Resource Review

1. List at least three renewable resources.

____
____
____

2. List at least three nonrenewable resources.

____
____
____

3. What are the differences between renewable and nonrenewable resources?

____
____
____

4. As the population on Earth increases, what happens to the natural resources?

____
____
____
____

Name ____

Mineral Mining

Write **T** for **true** or **F** for **false** after each statement. If the statement is false, correct it in the space below the sentence.

1. An area of rock that contains a large amount of a mineral is called a mineral deposit. _____

2. Ores containing minerals that are not metals are called metallic ores.

   _____

3. Nonmetallic ores are materials containing minerals that are not metals.

   _____

4. Gems are minerals. _____

5. Choose a gem. Using a textbook or reference materials, find out more about it. (For example, where is it found on Earth? How is it mined? What are its uses?) In the space below, write a paragraph about the gem you chose.

____

____

____

____

____

____

Name ______________________________

Energy Resources

1. Name three fossil fuels.

____

____

____

2. Explain the origin of fossil fuel.

____

____

____

3. Name and describe three energy sources that are not fossil fuels.

__________ : ______________________________

____

__________ : ______________________________

____

__________ : ______________________________

____

Name ______________________________

Conservation Concepts

Conservation can make energy and natural resources last longer. In the space below, write a paragraph describing how we can conserve energy resources and other natural resources.

Name ____________________

How Old Is That Rock?

The law of superposition states that a layer of an undisturbed section of sedimentary rock is older than those above it and younger than those below it. Using this information, answer the following question: If a layer of granite is between two layers of sedimentary rock, which layer is the oldest and which is the youngest? Explain your answer.

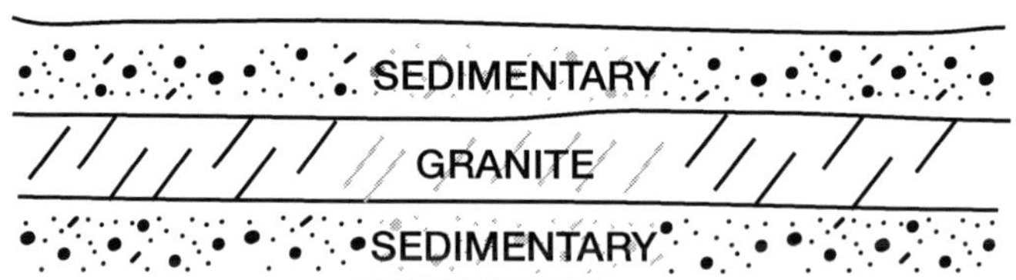

Name ____

Fossil Finder

1. If you are a paleontologist, what kind of work do you do?

____

____

____

2. Fossils are formed in several ways. Describe three of them.

____

____

____

____

____

____

3. How do molds and casts differ?

____

____

____

4. Draw a sketch of a fossil in the space below.

Name ______________________________

Past Perfect

1. Explain how fossils give clues about Earth's past.

____

____

____

2. What is a geologic column?

____

____

____

3. In the space below, draw a diagram of a geologic column.

Name ______________________________

Geologic Time

1. Name the four geologic eras.

_______________ _______________

_______________ _______________

2. Write the four geologic eras in order from the oldest to most recent.

Oldest **Most Recent**

3. Give an example of one creature that existed in each era.

Era	Creature

4. Draw a picture of one of the creatures listed above.

Name

Astrology Versus Astronomy

1. There are two groups of people who study the night sky. These people are astronomers and astrologers. In the space provided, explain the difference between an astronomer and an astrologer.

2. On the lines below, explain which study is more scientific—astronomy or astrology.

Name ______________________________

Around We Go!

1. Draw a diagram that illustrates what causes day and night.

2. Draw a diagram that illustrates what causes the change of the seasons.

Name ____

Mysterious Moon

Match each term to its definition. Write the correct letter on the line provided.

1. _____ satellite
2. _____ craters
3. _____ maria
4. _____ solar eclipse
5. _____ lunar eclipse

a. the smooth places on the Moon's surface

b. an object in space that revolves around a planet

c. occurs when Earth passes between the Moon and the Sun

d. bowl-shaped holes

e. occurs when the Moon passes between Earth and the Sun

6. Draw a diagram of what you think the surface of the moon looks like.

Name

Phase Craze

Draw a simple diagram that illustrates what each phase of the moon looks like.

new moon

first crescent

first quarter

first gibbous

full moon

last gibbous

third quarter

last crescent

Name ______________________________

Sun-sation!

Fill in each line with the word or phrase from the box that best completes the sentence.

star	photosphere	corona
150 million	sunspots	solar flares
hydrogen and helium		

1. At times there are bright areas in the Sun's atmosphere called ____________________.

2. The Sun is a ______________.

3. The Sun is _______________ miles from Earth.

4. _________________ are places on the Sun where the gases have cooled.

5. The Sun's surface is the ________________.

6. The Sun is made of the gases ____________________.

7. The faint white light around the edges of the Sun is called the ____________________.

Name ______________________________

Planet Power

The four inner planets of our solar system are listed in the box below. Write the name of the planet from the box that is described by each phrase.

Mercury	Venus	Earth	Mars

1. the smallest planet ________________
2. the fourth planet from the Sun ______________
3. the only planet to have oxygen in its atmosphere ________________
4. the closest planet to Earth ___________________
5. the red planet ___________________
6. the planet that can have a surface temperature as high as 430 degrees Celsius ___________________
7. the planet that is covered by yellow clouds __________________
8. the third planet from the Sun _____________________
9. the planet closest to the Sun ____________________
10. the second planet from the Sun ____________________

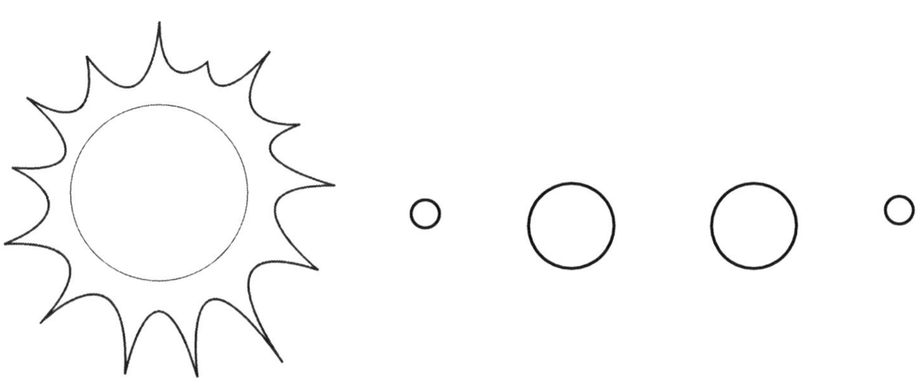

Name ______________________________

The Outer Planets

The five outer planets are listed below. Fill in the lines with facts about each planet.

Jupiter	Saturn	Uranus	Neptune	Pluto
1. It is the _____ planet from the Sun.	1. It is the _____ largest planet.	1. It is the _____ planet from the Sun.	1. It is the _____ planet from the Sun.	1. It is the _____ planet from the Sun.
2. It has ___ moon(s).	2. It is the _____ planet from the Sun.	2. It lies on its _____.	2. It has ___ moon(s).	2. It has ___ moons.
3. About ______ Earths could fit into it.	3. It has ___ moon(s).	3. It has ___ moon(s).	3. It is about the same size as ______.	3. It is about the same size as Earth's ______.

Name ______________________________

Our Solar System

Draw and label a diagram of the solar system including the Sun and the nine planets. Pay close attention to the order of the planets from the Sun and their size.

Name ____

Planet Personality

Many of the planets were named after Roman gods. Match the planet to the description of the god it was named after. Write the letter of the correct planet on the line provided.

1.	____ the ancient Roman messenger god	**a.**	Pluto
2.	____ the Roman goddess of love	**b.**	Mars
3.	____ the Roman god of war	**c.**	Neptune
4.	____ the king of the Roman	**d.**	Venus
5.	____ the Roman god of the harvest	**e.**	Saturn
6.	____ the Roman god of the heavens	**f.**	Uranus
7.	____ the Roman god of the sea	**g.**	Jupiter
8.	____ the Roman god of the dead	**h.**	Mercury

Name ____

Planet Quest

Choose one planet that you would like to know more about. Then answer the following.

Planet you chose:____________________

1. location ____
2. distance from the sun ________________________________
3. length of a year _____________________________________
4. diameter ____
5. average surface temperature ___________________________
6. major atmospheric gases ______________________________
7. number of moons _____________________________________
8. length of a day ______________________________________

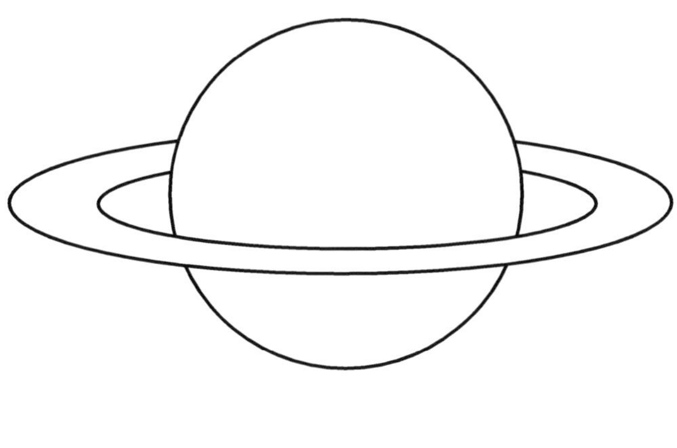

Name ______________________________

Don't Call Me Star

Answer the following in complete sentences.

1. What is the difference between a meteor and a meteorite?

2. Describe the materials that make up a comet, a meteor, and an asteroid.

3. Where are asteroids found?

4. Explain the travel pattern of comets.

Name

Where Did It Go?

Draw a diagram that illustrates a lunar eclipse and one that illustrates a solar eclipse. In each diagram include the positions of Earth, the Moon, and the Sun.

Lunar Eclipse:

Solar Eclipse:

Name ______________________________

Star Stuff

Fill in each line with the word or phrase from the box that best completes the sentence.

Proxima Centauri	white dwarf	black hole
dwarfs	quasars	magnitude
giants	radio telescope	super giants

1. Very large stars are called ______________________.
2. The star closest to Earth is ______________________.
3. The brightness of a star is called ______________________.
4. As the inside of a star collapses and the star shrinks, it becomes a ______________________.
5. Stars that swell and become large are ______________________.
6. Objects that are about the same size as stars, move very fast, and give out large amounts of energy are called ______________________.
7. The radio waves that are emitted by a quasar can be detected using a ______________________.
8. Small stars are called ______________________.
9. When some giant stars collapse, they form a ______________________.

Name ______________________________

Life Cycle of a Star

On the lines provided, write the stages of a star's life in order from beginning to end. Use the words in the box below.

nebula	red giant	black dwarf
star	white dwarf	protostar

1. ______________________
2. ______________________
3. ______________________
4. ______________________
5. ______________________
6. ______________________

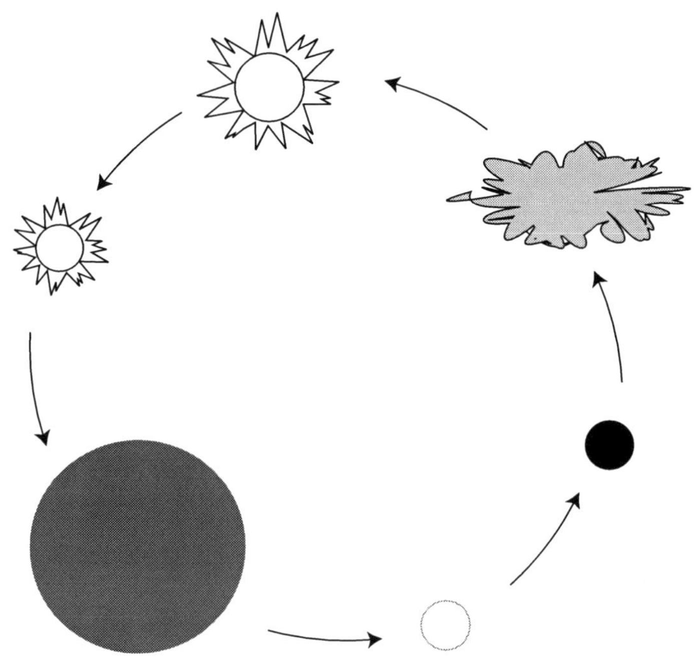

Name ______________________________

Constellations

In the space provided, answer the questions using complete sentences.

1. Can you name one constellation with which you are familiar? Which is it?

____

____

2. Why do constellations seem to move across the sky?

____

____

____

____

3. Why do we see different constellations at different times of the year?

____

____

____

____

Name ______________________________

Galaxies

Answer each of the following in the space provided.

1. There are four types of galaxies. Name and describe each.

   a. ______________________________

   ____

   b. ______________________________

   ____

   c. ______________________________

   ____

   d. ______________________________

   ____

2. Draw a sketch of each type of galaxy.

   a.

   b.

   c.

   d.

Name ____

Telescopes

There are two types of telescopes that are commonly used—the refracting telescope and the reflecting telescope. Use the Venn diagram below to compare and contrast the two types of telescopes. Use a textbook or reference materials, if needed.

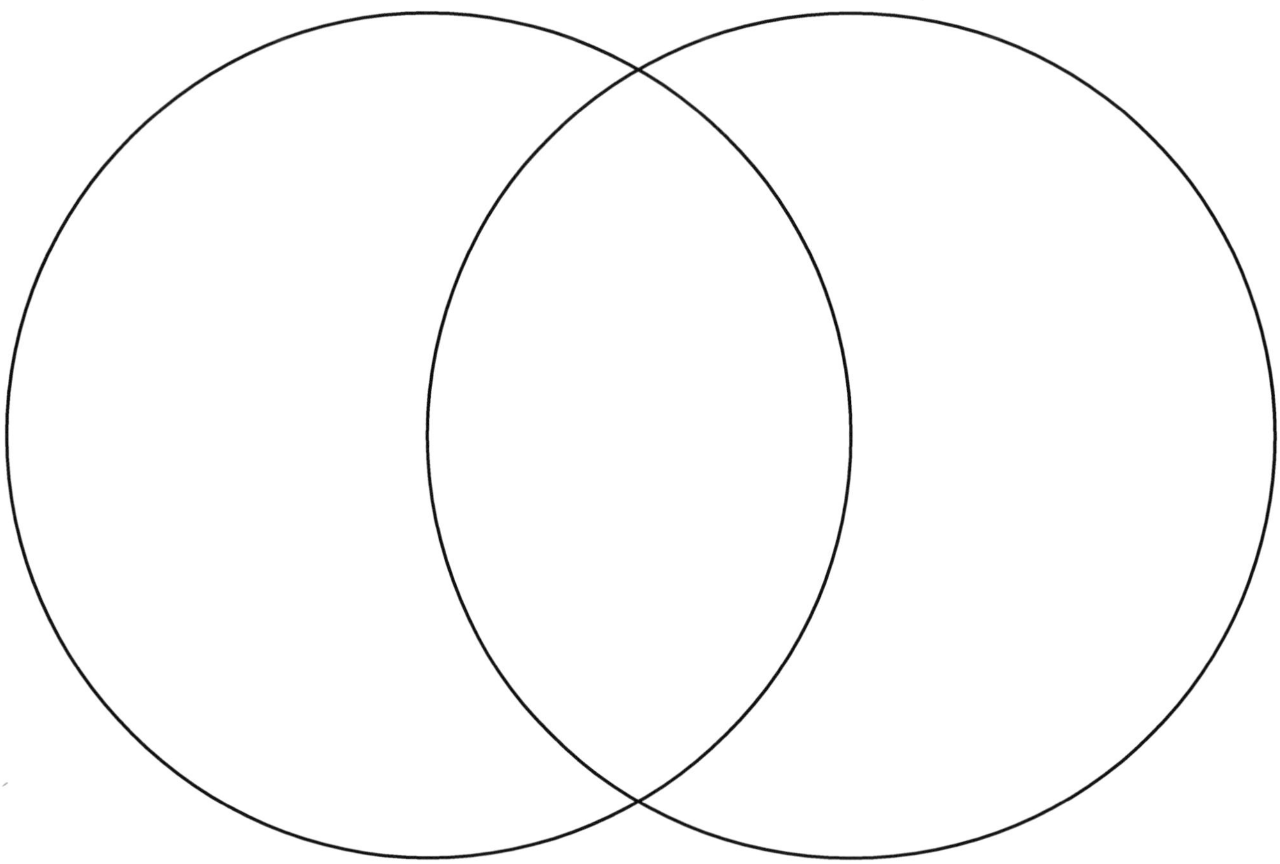

Name ______________________________

The Space Shuttle

In the space below, write a paragraph describing what you know about NASA's Space Shuttle program.

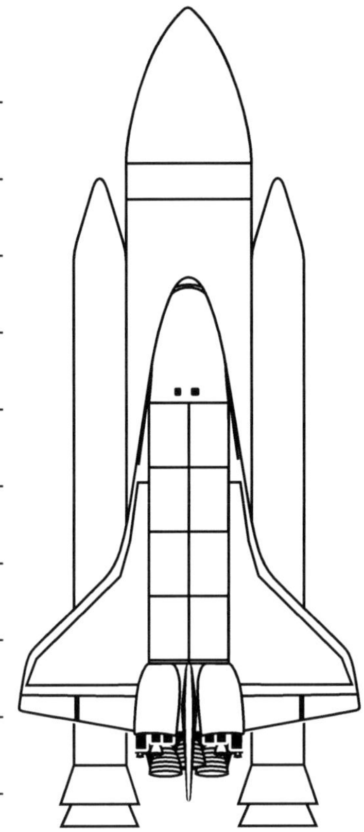

What search terms could you use in an Internet search engine to find out more information? List six terms below.

Name ______________________________

The Ocean

Provide the following information about the ocean.

1. Name the three main parts of the ocean bottom.

   a. ______________________________

   b. ______________________________

   c. ______________________________

2. Describe each of the three main parts of the ocean bottom.

   a. ______________________________

   b. ______________________________

   c. ______________________________

3. Draw a diagram of the ocean bottom, and label the three main parts.

Name ______________________________

Salts and Sediments

Provide the following information about the ocean.

1. Explain how the ocean becomes salty.

____

____

____

2. Draw a diagram that explains how the ocean becomes salty.

3. Briefly explain how each of the following arrives at the ocean floor.

   a. mud and clay from rivers: ______________________________

   ____

   b. animal remains: ______________________________

   ____

   c. plant remains: ______________________________

   ____

   d. rock particles: ______________________________

   ____

Name ______________________________

Ocean Movements

Fill in each line with the word from the box that best completes the sentence.

Gulf Stream	current	westerlies	California Current	trade winds

1. Water that moves in a certain direction is a ____________________.
2. The winds that blow from east to west are called ____________________.
3. The winds that blow from west to east are called ____________________.
4. The ____________________ is a current in the Atlantic Ocean.
5. The ____________________ is a current in the Pacific Ocean.

upwelling	salinity	temperature	nutrients	density

6. Cold, nutrient-rich water comes up from the bottom of the ocean to the surface through a process called ____________________.
7. The main reason that vertical currents happen in the ocean is due to a difference of ____________________.
8. A density difference can be caused by a change in ____________________ or ____________________.
9. Upwelling brings ____________________ to the surface. This provides food for the ocean creatures at the surface.

Name ____

Waves

Provide the following information about waves.

1. Explain how wind causes waves.

____

____

____

2. Draw a diagram that illustrates how wind causes waves.

3. Draw and label a diagram that shows the crest, trough, height, and length of a wave.

4. Explain what a tsunami is and how it occurs.

____

____

____

Name ________________________________

The Story of Erosion

Water, ice, wind, and people erode the land. Imagine that you are a rock being eroded. Write a story about what happens to you and where you might end up.

Name ______________________________

How Is Soil Developed?

Soil is formed over time by the weathering of rock. Soil is classified as immature and mature. **Immature soil** has a layer of topsoil and a layer of rock fragments. **Mature soil** has three layers: topsoil, subsoil, and a layer of rock fragments. Answer the following to explain the formation of soil.

1. Explain the characteristics of each of the following:

   a. topsoil ______________________________

   ____

   b. subsoil ______________________________

   ____

   c. rock fragments ______________________________

   ____

2. Draw and label a diagram of mature soil that illustrates all three of its layers.

3. Explain how weather (wind, snow, rain, and so forth) helps to develop soil.

   ____

   ____

   ____

   ____

Name ____________________

Inside the Earth

Answer each of the following.

1. List as many characteristics for each of the terms as you can.

   a. crust: ____________________

   ____

   ____

   b. mantle: ____________________

   ____

   ____

   c. core: ____________________

   ____

   ____

2. What is happening at the mid-ocean ridge?

   ____

   ____

   ____

   ____

3. How many plates is the earth divided into?

   ____

Name ______________________________

Earthquakes

Match each term to its definition. Write the letter of the correct answer on the line provided.

1. _____ fault
2. _____ primary waves
3. _____ secondary waves
4. _____ surface waves
5. _____ seismograph

a. an instrument that records earthquake waves

b. waves caused by back and forth vibrations in the rock

c. the result of the movement of rock along either side of a crack in Earth's crust

d. waves that cause the surface of Earth to move in ripples

e. waves that move in an up-and-down motion

Now use three of the words from above in sentences of your own.

6. ______________________________

____

7. ______________________________

____

8. ______________________________

____

Name ______________________________

Volcano!

Provide the following information about volcanoes.

1. Draw a diagram of a volcano, and label the vent, cone, and crater.

2. Name and describe each of the three kinds of volcanoes.

a. ______________________________
____
____

b. ______________________________
____
____

c. ______________________________
____
____

Name ____

Way Up There!

1. There are three types of mountains: folded, faulted, and domed. Describe how each of the three types of mountains is formed.

   a. folded: ____

   ____

   ____

   b. faulted: ____

   ____

   ____

   c. domed: ____

   ____

   ____

2. What is a plateau? ____

   ____

   ____

3. How is a plateau formed? ____

   ____

   ____

Name ____

Up, Up, and Away!

Earth's atmosphere is divided into five layers: troposphere, stratosphere, mesosphere, thermosphere, and exosphere. List three characteristics of each layer of the atmosphere. Use a textbook or reference materials, if needed.

1. troposphere:

   a. ____

   b. ____

   c. ____

2. stratosphere:

   a. ____

   b. ____

   c. ____

3. mesosphere:

   a. ____

   b. ____

   c. ____

4. thermosphere:

   a. ____

   b. ____

   c. ____

5. exosphere:

   a. ____

   b. ____

   c. ____

Name ___

Staying Cool Under Pressure?

Air pressure and temperature are two factors that affect weather. **Air pressure** is the force caused by the weight of the air. **Temperature** is the warmth or coldness of the air. Answer the following using complete sentences.

1. Air pressure is affected by two things. What are these two factors?

___
___
___

2. Do air particles prefer to be in an area of high pressure or in an area of low pressure? Explain in a complete sentence.

___
___
___

3. What instrument is used to measure air pressure? Explain how it works.

___
___
___

4. Scientists believe that three things happen to sunlight as it enters Earth's atmosphere. Explain these three things.

___
___
___

5. The areas of Earth closest to the equator have the warmest weather, while places far from the equator have colder weather. This is caused by the angle at which the Sun's rays hit Earth. In the space below, draw a diagram that illustrates the way the Sun's rays hit Earth.

Name ____________________

Windy Weather

In Earth's upper troposphere, warm and cool air are constantly changing places. This air movement and the rotation of Earth create large, circular wind systems close to Earth. These winds are referred to as the **jet stream.** In the lower troposphere, there are local winds. These are the winds that we feel in our daily weather.

Answer the following.

1. True or false: The winds in the jet stream flow from west toward the east.

   ____

2. Why are planes heading east helped by the jet stream?

   ____

3. True or false: The wind speeds in the jet stream always remain constant.

   ____

4. Explain the difference between a sea breeze and a land breeze.

   ____

   ____

   ____

   ____

   ____

5. Explain how large geological features (mountains, oceans, and so forth) affect winds.

   ____

   ____

   ____

Name ____

Cloud Creations

There are three types of clouds: cirrus, cumulus, and stratus. In the boxes below, draw a sketch of each type of cloud. List three facts about each on the lines provided.

1. stratus:

a. ______________________________

____

b. ______________________________

____

c. ______________________________

____

2. cumulus:

a. ______________________________

____

b. ______________________________

____

c. ______________________________

____

3. cirrus:

a. ______________________________

____

b. ______________________________

____

c. ______________________________

____

Name ______________________________

Moisture Mind-Bender

Moisture evaporates from Earth's surface and condenses. As it rises in the atmosphere, it forms clouds. The moisture that returns to Earth's surface from this process is known as **precipitation.** There are four types of common precipitation.

List the four major types of precipitation and explain the formation of each.

1. ______________________

Explain: ______________________________

____

____

2. ______________________

Explain: ______________________________

____

____

3. ______________________

Explain: ______________________________

____

____

4. ______________________

Explain: ______________________________

____

____

Name ______________________________

Fabulous Fronts

Weather changes are generally related to air masses. Air masses are large bodies of air that cover land and water. Air masses take on the characteristics of the areas over which they form. The boundary at which two air masses meet is called a **front.** There are three basic types of fronts.

Answer the following.

1. Air masses forming over land are called__________________, and those forming over water are ____________________. Air masses forming at high, colder latitudes are called_________________, and those forming at lower, warmer latitudes are called__________________.

2. What are the three types of fronts?

____

____

3. What type of front is formed when a warm air mass meets a cold air mass? Explain what happens to the weather as this occurs.

____

____

____

4. What type of front is formed when a cold air mass meets a warm air mass? Explain what happens to the weather as this occurs.

____

____

____

Name ______________________________

Why Forecast the Weather?

People have always been interested in the weather. Since 300 B.C.E., scientists have worked to make weather forecasting a more reliable science. There are many reasons why weather is important to people. Think of some reasons why forecasting the weather would be important to you. In the space provided, write a paragraph explaining your reasons.

Name ____________________

Tools of the Trade

Many different types of tools are used to assist meteorologists in forecasting the weather. Below, match each tool to the correct use or function of the tool. Write the letter of the correct answer on the line provided.

1. ____ observation stations

2. ____ weather balloons

3. ____ polar orbiting satellites

4. ____ geostationary satellites

a. These tools maintain an altitude of 500–900 miles above Earth's surface. They orbit the poles and view the different parts of the planet.

b. There are more than 3,500 of these around the world. They use instruments to measure many conditions, such as air temperature, air pressure, wind direction, and wind speed.

c. These tools contain instruments that measure air temperature, air pressure, and humidity at different levels in the atmosphere.

d. These tools maintain an altitude of 22,300 miles above Earth's surface and remain focused on one portion of Earth.

Name ______________________________

Weather Maps

There are two different kinds of weather maps: **surface maps** and **forecast maps.**

1. In the boxes below, list the things that each type of weather map can tell us about the weather.

Surface Weather Maps	Forecast Weather Maps

2. Using a Venn diagram, compare and contrast the two types of weather maps.

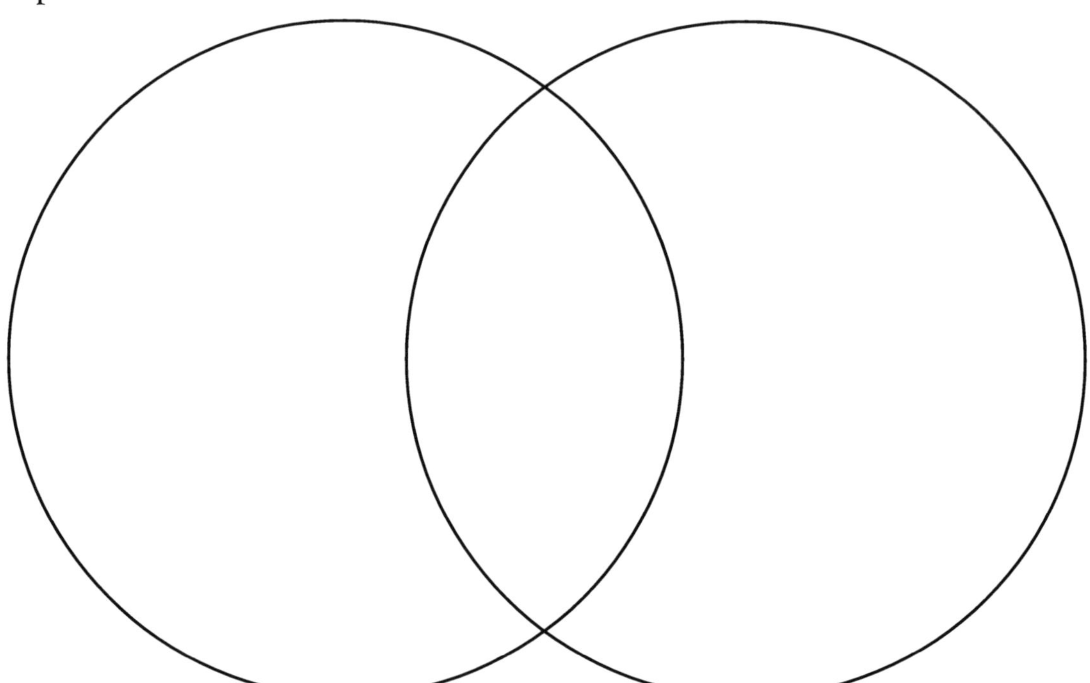

Name ______________________________

Natural Versus Synthetic

Some of the items we use every day come from natural materials. Other items are synthetic. A **natural** material comes directly from Earth or is made from materials found on Earth. A **synthetic** item is made by people and is not made of natural materials. Look at the list of items below. Think about what appears in nature, and label each item as natural or synthetic.

1. lumber ____________________

2. tires ____________________

3. vegetables ____________________

4. steak ____________________

5. polyester fabric ____________________

6. nylon fabric ____________________

7. cotton fabric ____________________

8. stereos ____________________

9. cupcakes ____________________

Name ___

Which Is Better?

Think about natural versus synthetic items. Which do you believe are the best items to use most frequently? Choose a side, and write a persuasive paragraph on the lines below. A **persuasive paragraph** is one that attempts to change someone's mind about an issue.

Name

Natural and Synthetic Do Battle

Draw a comic strip that features characters that represent natural and synthetic. You need to choose which character will be the bad guy and which character will be the hero. The two will battle about which is better for the environment and Earth. In the boxes below, draw your comic strip.

1.	2.
3.	4.

Name ________________________________

Science Versus Technology

Use a science textbook, dictionary, or other reference materials to help you find the difference between science and technology. In the space below, list at least three concepts or studies that are science. Then write a concept in technology that would be related.

Example:

Science:	Technology:
A physicist studies wind.	Car designs change to be more aerodynamic.

Science:	**Technology:**
____________________	____________________
____________________	____________________
____________________	____________________
____________________	____________________
____________________	____________________
____________________	____________________
____________________	____________________
____________________	____________________

Name

How Can Technology Help?

The creation of plastic was a great use of technology. Use the Internet, a textbook, or other reference materials to find information about the many uses of plastic. In the space below, write a paragraph that tells why you think one use of plastic is the most beneficial and explain why.

Name

Science and Technology Matchup

Match the technology to the science from which it was derived. (For example: The scientific study of human kidneys led to the technology of kidney machines for dialysis.) Write the letter of the correct technology on the line provided.

1. ____ the scientific study of trees
2. ____ the study of gravity and the rotation of the planets
3. ____ the scientific study of human health
4. ____ the scientific study of plants

a. the technology of improved crop varieties

b. the technology of paper production

c. the technology of artificial satellites

d. the technology of medicines

Name ____

Health Fact or Fiction?

Listed below are several pairs of health facts and health superstitions. Based on what you have learned about good health, decide which statement in each pair is the health superstition and circle it.

1. **a.** Men have one rib fewer than women do.
   b. Men and women have the same number of ribs.

2. **a.** You can catch a cold from being cold.
   b. Colds are caused by viruses and are passed on from person to person through contact.

3. **a.** Fish contains omega fatty acids that are healthy for your heart.
   b. Fish is brain food.

4. **a.** Handling toads causes warts on your skin.
   b. Warts are caused by viruses.

5. **a.** Alcohol impairs your judgment and coordination.
   b. Alcohol is a stimulant to your body.

Name

The Magic Potion for Good Health?

Think of all the commercials that you see that advertise health and beauty products that will improve your health and your appearance. These products include items such as herbal blends and wrinkle cream. These items do not always do what the commercials claim.

In the space below, design a magazine advertisement for one of these items.

Name ______________________________

You Are What You Eat

As you grow up, it is important that you learn about healthy eating. The first step is learning how to read the labels on food. Use the label below to answer each question on the line provided.

Nutrition Facts

Serving Size 1 Cup
Servings per container About 2

Amount per serving	
Calories 210	**Calories from Fat 20**
Total Fat 2g	**3%**
Saturated Fat 1g	**5%**
Cholesterol 5mg	**2%**
Sodium 1020mg	**43%**
Total Carbohydrate 41g	**14%**
Fiber 3g	**12%**
Sugars 14g	
Protein 7g	
Vitamin A 10%	**Vitamin C 0%**
Calcium 4%	**Iron 10%**

1. How many servings are in this can? __________
2. If you ate the whole can, how many calories would you consume? __________
3. If you ate only one serving, how many grams of fat would you consume? __________
4. How many grams of sugars are in one serving? __________
5. What percentage of daily recommended fiber is one serving? __________
6. What is the percentage of daily recommended calcium in one serving?
   ____
7. What percentage of daily recommended sodium would you be consuming if you ate the whole can? __________

Name ____

Eating Healthy

It is important that you eat foods from each of the five food groups each day. The five food groups are the dairy group, the meat and protein group, the vegetable group, the fruit group, and the bread and cereal group. Also included in the food pyramid are the fats and sweets grouping. Use the food pyramid below to develop a healthy menu for one day. Be sure to include the proper number of servings from each food group.

Breakfast: ____________________________
____
____
____

Snack: ____________________________
____
____
____

Lunch: ____________________________
____
____
____

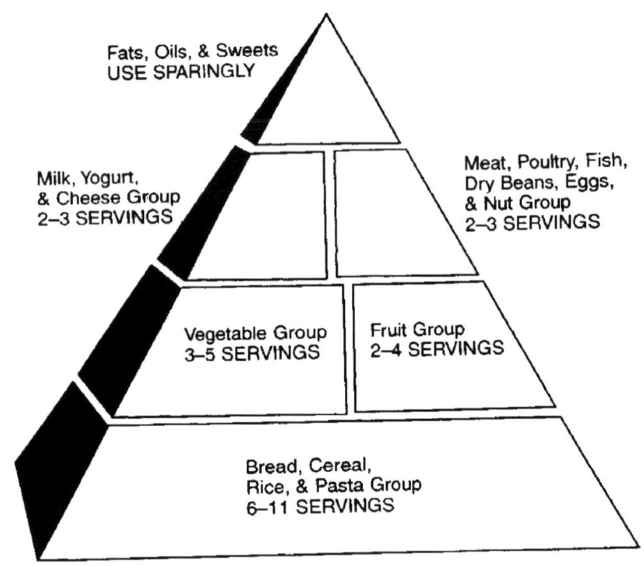

Snack: ____________________________
____
____

Dinner: ____________________________
____
____

Snack: ____________________________
____
____

Name ________________________________

Is Your Eating Healthy?

Compare the foods you ate yesterday with the food pyramid below.

List all the food you ate yesterday.

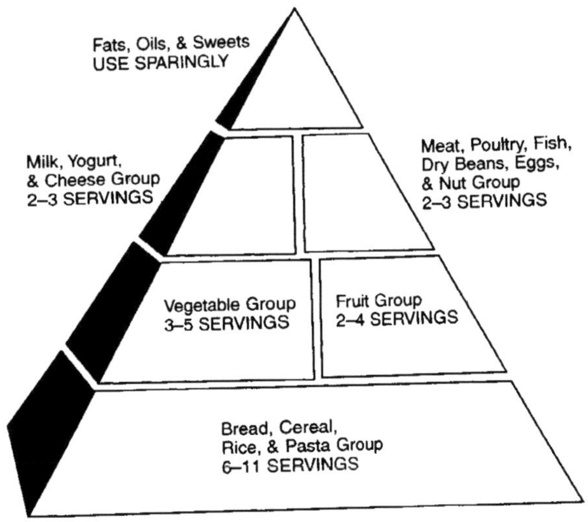

In a brief paragraph, explain how you could be a healthier eater, or why you think you are already a healthy eater.

Name

The Dangers of Drug Abuse

People are exposed to the dangers of illegal drug use and the abuse of other drugs every day. Think of posters that might be hung somewhere in your school, or of television commercials that you have seen. These commercials or posters give reasons to stay away from drugs. They explain the dangers of using or abusing drugs. In the space below, it is your turn to send a message to your friends and classmates about the dangers of drug use. Design a poster that you would like to hang in your school.

Name

Choices and Alcohol

Think of all of the reasons why people drink alcohol. In the space below, write a short story about a person who drank alcohol when he or she should not have. Describe the outcome of that person's actions.

Name

Fighting Tobacco Use

Think of all of the reasons why people should not use tobacco of any kind. In the boxes below, write and illustrate a comic strip about a superhero who fights the use of tobacco.

1.	2.
3.	4.

Name ____

Renewable or Nonrenewable?

Natural resources are divided into two categories: renewable and nonrenewable. **Renewable resources** are those that can be replaced by nature once people have used them. **Nonrenewable resources** are those that cannot be replaced. In the space below, answer the following.

1. Compare renewable and nonrenewable resources.

____

____

____

____

2. List three renewable resources.

a. ______________________

b. ______________________

c. ______________________

3. List three nonrenewable resources.

a. ______________________

b. ______________________

c. ______________________

Name ______________________________

Energy Resources

We use energy every day to drive cars, to light classrooms, or to run factories. There are many ways to get the energy that we need to live our lives. Listed below are some of the ways that we get energy. Use a textbook, the Internet, or reference materials to explain each.

1. fossil fuels: ______________________________

____

____

2. solar energy: ______________________________

____

____

3. water energy: ______________________________

____

____

4. geothermal energy: ______________________________

____

____

5. nuclear energy: ______________________________

____

____

Name ______________________________

Living Resources

Plants and animals are our living resources. These are important to people in many ways. Use a textbook, the Internet, or research materials to find ways in which plant and animal resources are important to us. In the space below, list five ways that plants and animals are important to people.

Plants:	Animals:
a. ______________________	a. ______________________
b. ______________________	b. ______________________
c. ______________________	c. ______________________
d. ______________________	d. ______________________
e. ______________________	e. ______________________

Name ______________________________

Identifying Populations

A **population** is a group of living things that are of the same species and live in the same area. Think of the types of specific plants and animals that are in the area where you live. In the space below, list five populations of living things that are in your area.

1. ______________________________
2. ______________________________
3. ______________________________
4. ______________________________
5. ______________________________

Now choose one of the populations you listed and draw a picture of its habitat. Maybe more than one of your populations live in the same habitat!

Name ______________________________

Change Is Good

An adaptation is a change in a species. This change helps the species survive in its environment. The change can be physical. The change can also be behavioral, or affecting the way an organism acts. Use the Internet, your text, or other reference materials to research an example of adaptation. Then, on the lines below, explain how that adaptation helped the species survive.

____

____

____

____

____

____

____

Draw a diagram that illustrates the adaptation you described above.

Name ______________________________

Technology Fun

Technology is making constant advancements. You are surrounded by it. Think of something that you use every day that was invented with the use of technology. Some examples of such inventions are the VCR, DVD player, computers, and the microwave oven. Choose one of the above mentioned items, or something different, and use a textbook, the Internet, or reference materials to answer the questions below.

Name of item you chose: ______________________

When was the item invented? ______________________________

How do you personally feel about the invention of this item?

____

____

____

____

What type of science, such as physics, biology, and so forth, do you think is connected to the technology that was used to invent this item?

____

____

Name

Good or Bad?

Some people think that technology has helped society. Some people think that technology has hurt society. Think about this, and decide how you feel about technology. Do you think it helps or hurts society? In the space below, write a paragraph explaining why you think it helps or hurts.

Name ____________________

Zoologist

A **zoologist** is a type of scientist. Use a textbook or reference materials to answer the questions below.

1. Is zoology a branch of biology or chemistry?

 ____

2. What does a zoologist study?

 ____

3. If you were a zoologist, where would you work?

 ____

4. Do you think you would like to be a zoologist? Why or why not?

 ____

 ____

 ____

5. Draw an illustration of something you would do if you were a zoologist.

Name ____________________

Botanist

A **botanist** is a type of scientist. Use a textbook or reference materials to answer the questions below.

1. Is botany a branch of physics or biology?

 ____

2. What does a botanist study?

 ____

3. If you were a botanist, where would you work?

 ____

4. Do you think you would like to be a botanist? Why or why not?

 ____

 ____

 ____

5. Draw an illustration of something you would do if you were a botanist.

Name ____________________

Chemist

A **chemist** is a type of scientist. Use a textbook or reference materials to answer the following questions.

1. Is chemistry a branch of biology or physical science?

   ____

2. What does a chemist study?

   ____

3. If you were a chemist, where would you work?

   ____

4. Do you think you would like to be a chemist? Why or why not?

   ____

   ____

   ____

5. Draw an illustration of something you would do if you were a chemist.

Name ____________________

Physicist

A **physicist** is a type of scientist. Use a textbook or reference materials to answer the questions below.

1. Is physics a branch of earth science or physical science?

   ____

2. What does a physicist study?

   ____

3. If you were a physicist, where would you work?

   ____

4. Do you think you would like to be a physicist? Why or why not?

   ____

   ____

   ____

5. Draw an illustration of something you would do if you were a physicist.

Name ______________________________

Geologist

A **geologist** is a type of scientist. Use a textbook or reference materials to answer the questions below.

1. Is geology a branch of earth science or physical science?

____

2. What does a geologist study?

____

3. If you were a geologist, where would you work?

____

4. Do you think you would like to be a geologist? Why or why not?

____

____

____

5. Draw an illustration of something you would do if you were a geologist.

Name ______________________________

Astronomer

An **astronomer** is a type of scientist. Use a textbook or reference materials to answer the questions below.

1. Is astronomy a branch of earth science or biology?

   ____

2. What does an astronomer study?

   ____

3. If you were an astronomer, where would you work?

   ____

4. Do you think you would like to be an astronomer? Why or why not?

   ____

   ____

   ____

5. Draw an illustration of something you would do if you were an astronomer.

Name

Marine Biologist

A **marine biologist** is a type of scientist. Use a textbook or reference materials to answer the questions below.

1. Is marine biology a branch of earth science or biology?

2. What does a marine biologist study?

3. If you were a marine biologist, where would you work?

4. Do you think you would like to be a marine biologist? Why or why not?

5. Draw an illustration of something you would do if you were a marine biologist.

Name ____________________

Oceanographer

An **oceanographer** is a type of scientist. Use a textbook or reference materials to answer the questions below.

1. Is oceanography a branch of biology or earth science?

   ____

2. What does an oceanographer study?

   ____

3. If you were an oceanographer, where would you work?

   ____

4. Do you think you would like to be an oceanographer? Why or why not?

   ____

   ____

   ____

5. Draw an illustration of something you would do if you were an oceanographer.

Name ______________________________

Meteorologist

A **meteorologist** is a type of scientist. Use a textbook or reference materials to answer the questions below.

1. Is meteorology a branch of physical science or earth science?

   ____

2. What does a meteorologist study?

   ____

3. If you were a meteorologist, where would you work?

   ____

4. Do you think you would like to be a meteorologist? Why or why not?

   ____

   ____

   ____

5. Draw an illustration of something you would do if you were a meteorologist.

Name ______________________________

Isaac Newton

Use a textbook or reference materials to answer the questions below.

1. In 1687, Newton published three rules about force and motion. What are these rules named?

____

____

2. What are the three laws of motion?

____

____

____

____

____

____

____

____

____

____

____

____

____

Name

Nicolaus Copernicus

Use a textbook or reference materials to answer the following.

1. Copernicus was from which country?

____

2. When was Copernicus born, and when did he die?

____

____

3. Explain the theory that Copernicus had regarding the solar system.

____

____

____

____

4. Was Copernicus' theory widely accepted? Why or why not?

____

____

____

____

____

Name ____________________

Thomas Edison

Use a textbook or reference materials to answer the following.

1. Where was Edison born? When was Edison born?

____

____

____

2. Edison invented many things. List three things that he invented.

____

____

____

3. Which of Edison's inventions is the most important to your life? Explain.

____

____

____

____

____

____

____

____

____

Name

Albert Einstein

Use a textbook or reference materials to answer the questions below.

1. Where and in what year was Einstein born?

2. What is Einstein's theory of relativity?

3. Was Einstein a good student? Explain.

Name ______________________________

Benjamin Franklin

Use a textbook or reference materials to answer the following.

1. Explain the experiment that Franklin used to prove that lightning is electricity.

____

____

____

____

2. Draw an illustration of the experiment described above.

Name ______________________________

Galileo Galilei

Use a textbook or reference materials to answer the questions below.

1. Galileo was one of the first people to use what instrument?

____

____

2. What was his first discovery using a telescope?

____

____

3. What other things did Galileo discover using a telescope?

____

____

____

____

4. In addition to astronomy, Galileo made important discoveries in which scientific field?

____

____

Name ____________________

Niels Bohr and Ernest Rutherford

Use a textbook or reference materials to answer the following.

1. What did Bohr and Rutherford describe?

____

____

2. Draw an illustration of what Bohr and Rutherford first described.

Name ______________________________

Wilhelm Roentgen

Use a textbook or reference materials to answer the questions below.

1. What did Roentgen discover in 1895?

____

2. Roentgen was awarded what prize for his discovery of the X ray?

____

3. What part of his wife did Roentgen X-ray?

____

4. What are X rays used for?

____

____

5. Have you ever been X-rayed? If so, describe what it was like. If not, describe what you think it is like.

____

____

____

____

____

Name ____________________

Dmitry Mendeleyev

Use a textbook or reference materials to answer the following.

1. Mendeleyev was a chemist from what country?

____

2. During what years did he live?

____

3. Explain Mendeleyev's major contribution to science in reference to the periodic table of the elements.

____

____

____

____

____

4. Mendeleyev predicted the properties of elements that were unknown at the time. What three new elements proved his theory during his lifetime?

____

____

____

Name ______________________________

William Herschel

Use a textbook or reference materials to answer the following.

1. Before William Herschel became interested in astronomy, he was a

   ______________________________.

2. What planet did William Herschel discover?

   ____

3. In what year did he discover this planet?

   ____

4. How did he discover the planet?

   ____

   ____

   ____

   ____

   ____

   ____

Answer Key

SCIENCE AS INQUIRY

Page 1: Classmate Clues

1. Blue 5, Green 3, Brown 6, Hazel 2
2. brown
3. Brown is more frequent because more children have at least one parent with brown eyes, and they pass on the dominant gene to their children.

Page 2: Survey Superstars

Answers will vary. ***Sample answers:***

1. I would like to know students' favorite food.
2. I will hand out a survey.
3. I will organize the information in a table.
4. I think most students' favorite food is pizza.

Page 3: Conducting a Survey

Answers will vary. ***Sample answers:***

1. A survey is designed to find out information about a group.
2. I would hand out a survey form and ask students to mark answers.
3. I would set up a data table and maybe a bar graph.

Page 4: Survey Solutions

1. 15
2. 6
3. 9
4. 6
5. 1
6. 8
7. brown; answers will vary.
8. brown; answers will vary.
9. Answers will vary.

Page 5: Balancing Act

1. 512 g
2. 237 g
3. 91 g
4. 164 g
5. 346 g

Page 6: Measuring the Little Things

1. 4.2 cm
2. 2.4 cm
3. 5.7 cm
4. .4 cm
5. 8.9 cm
6. 7 mm
7. 30 mm
8. 16 mm
9. 2 mm
10. 23 mm

Page 7: Lots of Liquids

1. 24 mL
2. 3 mL
3. 34 mL
4. 13 mL

Page 8: Temperature Time!

1. 19 °C
2. 10 °C
3. 0 °C
4. 117 °C

Page 9: Quick Conversions

1. 2 cm
2. 4 m
3. 4 km
4. 200 mm
5. 6,000 m
6. 400 cm

Page 10: Magic Measuring

Answers will vary. ***Sample answers:***

1. 120 cm
2. 30 cm
3. 5 mm
4. 15 mm
5. 250 mL
6. 2 g
7. 24 g
8. 7 °C
9. 14 °C

Page 11: Compare and Contrast

Answers will vary. ***Sample answers:***

Flower	Both	Tree
Has floppy stem	Have leaves	Has a woody stem
Has bright yellow petals	Are green	Has branches

Page 12: Prediction Perfection

Answers will vary. ***Sample answers:***

Observation	Prediction
Clouds are gathering.	It will rain.
The floor is freshly waxed.	Someone will slip.
It snowed yesterday.	Students will wear boots to school.

Page 13: Creative Classifying

Answers will vary. ***Sample answers:***

1. Lunch bag
   3 books
   2 folders
   4 pencils
   1 pen
2. 4 groups
3. Lunch group—lunch bag
   Book group—3 books
   Folder group—2 folders
   Pen and pencil group—4 pencils and 1 pen
4. The books all go together, and the folders all go together. I write with pens and pencils, so they make up one group. Lunch is all by itself.

PHYSICAL SCIENCE

Page 14: Four Properties of Matter

1. mass, volume, density, weight
2. specific
3. mass
4. weight
5. volume
6. Density
7. matter

Page 15: Phases of Matter

1. d
2. f
3. b
4. e
5. c
6. g
7. a

Page 16: Phase Changes

1. Freezing point and melting point are both the same temperature. At freezing point, a liquid becomes a solid. At melting point, a solid becomes a liquid.
2. A gas changes to a liquid; examples will vary.
3. A liquid becomes a gas.
4. In both evaporation and boiling, a liquid becomes a gas. Evaporation takes place at the surface of the liquid, while boiling takes place throughout.

Page 17: Classes of Matter

Mixtures

A combination of two or more types of matter that can be physically separated; examples will vary (e.g., a peanut butter sandwich).

Compounds

A pure substance made of two or more elements that are chemically combined; examples will vary (e.g., table salt).

Elements

A substance that cannot be made into simpler substances; examples will vary (e.g., a gold ring).

1. mixture
2. elements
3. compound

Page 18: The Atom

1. the smallest particle of an element that still has the properties of that element
2. protons, neutrons, and electrons
3. negative
4. neutral
5. positive
6. protons and neutrons
7. around the nucleus
8. Diagrams will vary.

Page 19: Element or Molecule?

1. M
2. E
3. E
4. E
5. M
6. M
7. M
8. E
9. M
10. E
11. E
12. M
13. E
14. E
15. E

Page 20: Get to Know Me?

1. Cr
2. Hg
3. B
4. P
5. Pb
6. Se
7. Ar
8. Ta
9. Li
10. A
11. Fe
12. Ra
13. K
14. Rn
15. Ni
16. scandium
17. copper
18. germanium
19. platinum
20. iridium
21. bromine
22. cobalt
23. xenon
24. sulfur
25. helium
26. zinc
27. terbium
28. francium
29. magnesium
30. oxygen

Page 21: Who Am I?

1. lithium
2. rhenium
3. nickel
4. chlorine
5. tungsten
6. rubidium
7. calcium
8. calcium
9. rhenium
10. helium

Page 22: Protons, Neutrons, and Electrons

1. Ca, 40, 20, 20
2. copper, 29, 29, 35
3. W, 183.8, 74, 74
4. Sc, 44.9, 21, 24
5. potassium, 39.1, 19, 20
6. xenon, 131.30, 54, 54
7. 107.87, 47, 47, 61
8. iodine, 53, 53, 74
9. lead, 82, 82, 125
10. Hg, 200.59, 80, 121

Page 23: Speedy Scramble

1. acceleration
2. speed
3. velocity
4. distance
5. motion

6–10. Answers will vary. ***Sample answers:***

6. acceleration—increase in velocity given by $a = v/t$
7. speed—change in distance over time
8. velocity—a change in distance over a change in time in a given direction—$v = d/t$
9. distance—a given amount of space
10. motion—the movement of a body in space

Page 24: Newton Knowledge

1. second law
2. first law
3. third law

4–6. Answers will vary. ***Sample answers:***

4. A soccer ball will not move at kickoff until it is first kicked.
5. A soccer ball will accelerate in the direction it is kicked.
6. The foot of the person kicking the soccer ball will slow down or stop completely after contact with the ball.

Page 25: Motion Commotion

Answers will vary. ***Sample answers:***

1. Student slides down a slide. Second Law—the force of gravity causes the student to accelerate in the direction of the force.
2. Two students are on a seesaw. As one goes down, the other goes up. Third law—for every action, there is an equal but opposite reaction.
3. One student pulls another in a cart. When the first student lets go, the cart continues rolling for a time. First Law—a body in motion will continue in motion unless acted upon by an outside force.

Page 26: Are We There Yet?

1. 43.3 mi/hr
2. 20 mi/hr
3. 5 mi/hr

Page 27: Amazing Acceleration

1. 6 mph/s
2. 4 mph/s
3. .63 mph/s

Page 28: Finding the Force

1. 40 kgm/s^2
2. 24 kgm/s^2
3. 2 kgm/s^2

Page 29: Force Facts

1. weak interactions
2. friction
3. Gravitational force
4. magnetic force
5. Nuclear force
6. electrical force

Page 30: Forces of Liquids

1. c
2. b
3. a

4–6. Answers will vary.

Page 31: Work and Force

1. *effort force*—the force that is applied to do work
2. *resistance force*—the force that opposes motion
3. *work*—using force to move an object
4. 3,000 N × m
5. 2,500 N × m

Page 32: Waves and Sound

1. f	6. c
2. i	7. j
3. g	8. h
4. b	9. e
5. d	10. a

Page 33: Sound Waves

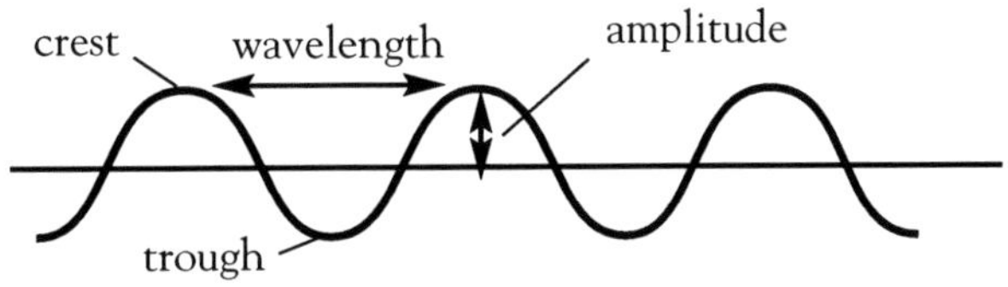

1. a measure of the greatest displacement of matter from its normal resting position
2. the distance from any point on a wave to the corresponding point on the next wave
3. the top of a wave
4. the bottom of the wave

Page 34: Reflection and Refraction

1. angle of reflection
2. incident ray
3. angle of refraction
4. the law of refraction
5. refraction
6. reflection
7. index of refraction
8. reflected ray

9–11. Answers will vary. ***Sample answers:***

9. Angle of refraction—the angle that refracted light makes with a surface
10. Refraction—the apparent bending of a light wave when it moves from one medium to another
11. Reflection—the bouncing back of a ray of light when it strikes a reflective surface, such as a mirror

Page 35: Light Logic

1. reflection	5. diffraction
2. concave	6. incandescence
3. light year	7. refraction
4. light rays	8. convex

Page 36: Looking at Light

1. When a ray of light is reflected from a smooth surface, the angle of reflection is equal to the angle of incidence.
2. Light is refracted so highly that white light breaks down into its components, producing all the colors in white light—the colors of the rainbow.
3. Both are a changing of direction of light rays. Reflection is the bouncing back of light rays. Refraction is the change of direction of light rays as they pass through a medium.
4. Answers will vary. ***Sample answers:***
   opaque—a wooden board
   transparent—a sheet of glass
   translucent—a piece of frosted plastic

Page 37: Let's Heat Things Up!

1. j
2. a
3. i
4. h
5. c
6. g
7. d
8. f
9. e
10. b

Page 38: Hot and Cold

1. addition of heat
2. loss of heat
3. addition of heat
4. addition of heat
5. addition of heat
6. Answers will vary. ***Sample answer:*** chilling hot food in the refrigerator
7. Answers will vary. ***Sample answer:*** using a blowtorch to cut metal

Page 39: Gas Laws

1. Boyle's law
2. Charles' law
3. Answers will vary. ***Sample answer:*** A diver rises slowly from the ocean floor to avoid the bends.
4. Answers will vary. ***Sample answer:*** Airplanes are kept very cool to avoid an uncomfortable increase in air volume.

Page 40: Energy for Engines

1. outside
2. Answers will vary. ***Sample answer:*** a steam engine
3. inside
4. Answers will vary. ***Sample answer:*** an automobile engine
5. a. The piston moves downward. This causes a mixture of fuel and air to be drawn into the cylinder.
   b. The intake valve closes, and the piston moves upward. The fuel-air mixture is compressed.
   c. A spark plug creates a spark that ignites the gas.
   d. The exhaust valve opens to let out the waste of the burning reaction.
6. A diesel engine does not have spark plugs.

Page 41: Magnet Magic

1. *magnet*—an object that picks up or attracts iron, nickel, and cobalt
2. *poles*—the ends of magnets
3. *lines of force*—lines around a magnet that show where the force fields are found
4. *magnetic field*—the space around a magnet in which there is a magnetic force
5. *electromagnet*—a magnet made with current electricity

Page 42: Scientists in the News

Answers will vary. ***Sample answers:***

1. Hans Christian Ørsted today announced that he had been able to deflect a compass needle from magnetic north when he passed an electric current near the needle. Said Ørsted, "This will open the door to finding the magical connection between electricity and magnetism."
2. Michael Faraday reports a new partnership with theorist James Clerk Maxwell in order to describe the unusual events Faraday has been reporting from his laboratory. Faraday's new idea of "lines of force" needed a mathematician and a theorist to make them come alive. Said Maxwell of the new partnership, "I had resolved to read no mathematics until I had read Faraday. But Faraday IS a mathematician."

Page 43: Magnetic Mania

1. Answers will vary. ***Sample answers:***
   Magnetic: iron nail, paper clip, iron filings, desk legs, other magnet
   Not magnetic: pencil, pen cap, desk top, blackboard
2. All magnetic things have some form of iron in them.

Page 44: Electricity Puzzler

1. Resistance
2. anode
3. insulator
4. watt
5. cathode
6. Voltage
7. conductor
8. kilowatt
9. Induction
10. electric circuit

Page 45: All About Electricity

1. b
2. c
3. a
4. b
5. b

Page 46: Circuit Secrets

Answers will vary. ***Sample answers:***

1. splits circuit through several resistors
2. sends entire circuit through several resistors
3. a chip or board that conducts electricity without wires
4. Most household wiring is parallel. If one resistor fails, the rest remain functioning.
5. Series circuits are often used in vehicles. This protects the car or truck against overheating or overloading.
6. Printed circuits are commonly used in computers.

Page 47: Super Scientists

Answers will vary. ***Sample answers:***

1. Charles Proteus Steinmetz invented alternating current.
2. Stephen Gray discovered that electricity could be conducted through copper wires.
3. Georg S. Ohm was the first to define the relationship between voltage, current, and resistance.

LIFE SCIENCE

Page 48: Common Sense

Answers will vary. ***Sample answers:***

1. Butterflies are insects and invertebrates. Snakes are reptiles and vertebrates. Butterflies undergo metamorphosis; snakes do not.
2. Dogs and cats are both mammals, and both give birth to live young. Dogs are pack animals, while cats tend to be more solitary.
3. Eagles are birds, and chipmunks are mammals. Both are vertebrates. An eagle is a predator that might eat a chipmunk.

Page 49: Similar Structures

Answers will vary. ***Sample answers:***

1. Both have wings and feathers; beaks and talons are different.
2. Both have fur and nurse young with milk; chipmunks are ground dwelling, while squirrels are tree dwelling.
3. Both have scales and fins; trout live in freshwater, and sharks live in salt water.

Page 50: Cell Game

1. g	5. d
2. h	6. c
3. a	7. f
4. e	8. b

Page 51: Calling All Cells!

1. cell wall	8. endoplasmic reticulum
2. cell membrane	9. ribosomes
3. protoplasm	10. mitochondria
4. nucleus	11. vacuoles
5. nuclear membrane	12. lysosome
6. chromosomes	13. chloroplast
7. nucleolus	

Page 52: Animal Cell

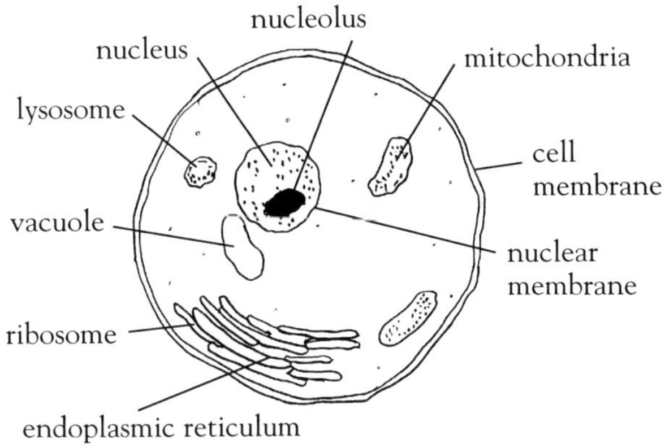

Page 53: Plant Cell

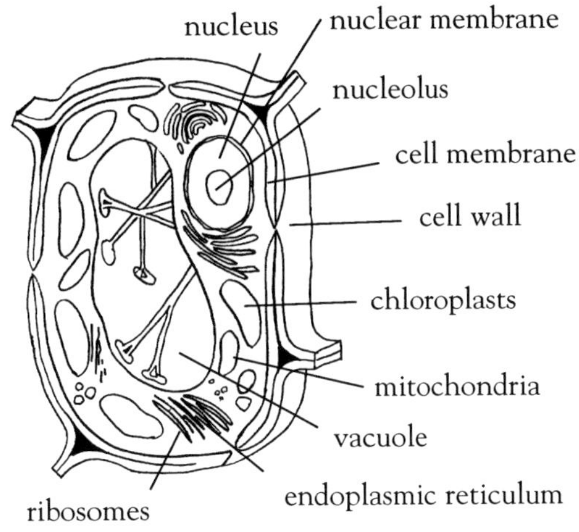

Page 54: Plant Versus Animal

Answers will vary. ***Sample answer:***

I am a plant cell, and I am more complicated because I have a cell wall, which you, as an animal cell, do not possess. I also have chloroplasts, which contain chlorophyll, and make it possible for me to make my own food.

No, I, the animal cell, am more complicated. I can take any shape I need to, while you are stuck behind that stiff cell wall. I have to take in nutrients and process them, while you make your own.

Page 55: Get Moving!

Answers will vary. ***Sample answers:***

living thing: frog
getting food: catching flies
movement: hopping or swimming
growing: complete metamorphosis
sensing and reacting: sensitive skin, large eyes that can see almost all directions
using food and removing wastes: cold blooded, slow metabolism. Respires through skin
development: complete metamorphosis—starts life as tadpole and grows legs, develops lungs, and can leave water for short periods.
reproduction: must return to water to breed and lay eggs

Page 56: King Me

Plant: rose, daisy, oak tree, grass
Animal: dog, frog, eel, ant
Protist: amoeba, euglena, sporozoan, paramecium
Fungi: puffball, mushroom, yeast, penicillin
Bacteria: spirillum, salmonella, E. coli, streptococcus

Page 57: Animal Kingdom

Sample answers:

Invertebrates: worm, snail, sea star, fly, spider, scorpion
Vertebrates: frog, snake, eagle, raccoon, trout

1. kingdom
2. phylum
3. class
4. order
5. family
6. genus
7. species

Page 58: Plant Kingdom

1. Nonvascular
2. vascular
3. dicots
4. parallel
5. monocot
6. network
7. gymnosperms
8. cone
9. conifers
10. Mosses

Page 59: Protist Puzzler

1. Plantlike protists make their own food.
2. Three features are chlorophyll, air sacs, and holdfasts.
3. Four groups are red algae, green algae, brown algae, and diatoms.
4. It moves by extending the edges of its cell membrane (pseudopodia).
5. An example is slime mold or water mold.

Page 60: Fungus Features

Answers will vary. ***Sample answers:***

1. Mold is a decomposer. It needs dead or dying plant or animal matter to survive.
2. Mold grows on old bread.
3. Molds break down animal and plant matter and release nutrients and water back into the environment.

Page 61: Behavior Basics

1. b
2. c
3. b
4. a
5. a
6. a

Page 62: Learned Behavior

Answers will vary. ***Sample answer:***

Dogs will come when you call them. When I call my dog, and she obeys, I give her a lot of attention and a little treat. This makes her want to come again when I call. This is called positive reinforcement. I can teach my dog a lot of things. I teach her to bring back a ball and walk right next to me when we are in the park. Every time, she gets good attention.

Page 63: Communication

Paragraphs will vary. ***Sample paragraph:***

Humans probably would not have survived without communication. We were able to tell one another when danger approached, or what was safe to eat. If we lost the ability to communicate today, it would be even worse, because we have lost a lot of instinctive behavior.

Page 64: Food Fun

1. b
2. e
3. h
4. f
5. g
6. c
7. a
8. d

Page 65: Digestion Question

1. fangs
2. gizzard
3. gall bladder
4. stomach
5. small intestine
6. Answers will vary. ***Sample answer:*** When we eat food, saliva starts breaking the food down in our mouths. We chew it to work the saliva in, and to make the particles smaller. We then swallow food, and it goes down our esophagus into our stomach. We have strong acids in our stomach that eat away at the food particles. Some of the nutrients are absorbed through the stomach wall and into the bloodstream. Most of the food, though, goes through the stomach into the small intestine, where it is absorbed. Food that is not absorbed goes into the large intestine, and finally the colon, where it is removed as solid waste.

Page 66: Respiration and Circulation

1. frog
2. jellyfish
3. insect

Page 67: Nervous System Scramble

Frog
chain of autonomic ganglia
cerebrum
optic lobe
cerebellum
medulla oblongata
cranial nerves
spinal cord
spinal nerves
brachial nerve
sciatic nerve

Earthworm
brain nerves
ventral nerve cord
ganglia

Grasshopper
brain
antennae
compound eye
nerves
ventral nerve cord
ganglia

Page 68: Plant Parts

1. The root absorbs water and minerals.
2. Xylem tissue, phloem, and root hairs are parts of the root.
3. It supports the plant and transports minerals and water throughout the plant.
4. The parts of a leaf are the petiole, midrib, vein, and blade.

Page 69: The Story of Photosynthesis
chlorophyll, chloroplast, roots, hydrogen, oxygen, carbon dioxide, sugars, oxygen

Page 70: The Oxygen–Carbon Dioxide Cycle
Descriptions and diagrams will vary. ***Sample answer:***
A plant absorbs carbon dioxide and water, and in the presence of sunlight, performs photosynthesis. A waste product of photosynthesis is oxygen, which the plants give off during the day. Animals and other organisms require oxygen for respiration. They, in turn, give off carbon dioxide, which is used by the plants.
Graphic should be a cycle and contain these elements:
Plant taking in CO_2 ~~~~~~~> Plant photosynthesizing in sunlight ~~~~~~~~~> Plant giving off oxygen ~~~~~~~~~~> Animal breathing in oxygen ~~~~~> Animal breathing out CO_2 ~~~~~~~> Plant taking in CO_2

Page 71: Plant Reproduction
1. T
2. T
3. F
4. F
5. T
6. T
7. F
8. T
9. T
10. T

Page 72: Life Cycles

Frog	**Butterfly**
egg	egg
embryo	larva
tadpole	pupa
legs develop	adult butterfly
adult frog	

Page 73: A Chip off the Old Block
Answers will vary. ***Sample answers:***
1. DNA is a genetic code. We need it to store and pass on genetic information.
2. Animal populations have some variations. Insects might be spotted or solid black. If conditions change in the environment, one of them might have a better chance for survival.
3. Animals reproduce to produce more of the same species; if organisms did not reproduce, they would become extinct.

Page 74: Heredity
1. j
2. b
3. h
4. d
5. e
6. i
7. f
8. c
9. g
10. a

Page 75: Mendel's Peas
1. Each pea plant parent has a pair of factors or a gene pair.
2. Some offspring were short.
3. All the plants were tall.
4. Answers will vary. ***Sample answer:***
   Mendel's peas showed that heredity is passed from parent plant to the offspring. The dominant genes are expressed in the young plant.

Page 76: Chromosomes
1. It is an alteration in genes or chromosomes.
2. Answers will vary. ***Sample answer:***
   Sickle cell anemia, in humans, is a genetic mutation.
3. They control the production of protein.

Page 77: Biotic and Abiotic
1. animals
2. biotic
3. abiotic
4. Ecology
5. Answers will vary. ***Sample answers:***
   Biotic Factors: Students, teacher, plant, fish in fishbowl, lunches in lunchboxes
   Abiotic Factors: chalkboard, pens, desks, temperature, amount of light

Page 78: Ecosystem Essay
Descriptive essays will vary. ***Sample essay:***
I visited the seashore. The abiotic features included the water and sand, and the biotic features were the birds, algae, insects, fish, sand dollars, and other life forms. In one pile of algae I found nine different kinds of insects. A biologist told me that they live their entire lives between one high tide and the next in the tangle of algae. The insects feed the small birds and other small animals, which are fed on by the seagulls and bigger animals.

Page 79: Ecosystem Equation
Answers will vary. ***Sample answer:***
Crab
community: rocky seashore
population: all the crabs in the tide pool
habitat: tide pool

Page 80: Habitat Match
1. d
2. e
3. c
4. b
5. h
6. f
7. g
8. a

Page 81: What Do I Eat?
1. an organism that can make its own food; example: wheat
2. an organism that feeds on producers; example: cow
3. an organism that eats only plants; example: horse
4. an organism that eats plants and animals; example: human
5. a meat-eating animal; example: wolf

6. an organism that feeds on dead organisms; example: mushroom

Page 82: Food Chain Fun
Diagrams will vary. ***Sample diagrams:***
1. cyanobacteria—small fish—heron—mold
2. acorn—wild pig—bobcat—bacteria
3. diatoms—snails—sea stars—bacteria

Page 83: Relationship Report
1. struggle among living things to get the proper amount of food, water, and energy (example: the competition between cultivated plants and weeds in a garden)
2. relationship that exists between predator and prey (example: an osprey and fish)
3. symbiotic relationship in which only one partner in the relationship benefits (example: In an abandoned anteater nest, a small antelope hides from predators. The anteater does not benefit from the relationship, but neither is it harmed. The antelope, however, benefits.)
4. symbiotic relationship that is helpful to both organisms (example: An Egyptian plover, a small bird, sits in the mouth of a Nile crocodile, and eats food particles that cling to the crocodile's teeth. The small bird gets a meal, and the crocodile gets a dental cleaning.)
5. relationship in which one organism benefits but the other is harmed (example: A tick attaches itself to the body of a deer, and sucks its blood. The deer is harmed, because it is losing blood, and possibly being infected with disease, but the tick benefits from the relationship.)

Page 84: Cycle Circles
Diagrams will vary.

Page 85: Biome Buster
1. biome that rims the Arctic Ocean around the North Pole and has a cold, dry climate
2. northernmost forest biome that contains conifers
3. forest biome that contains trees whose leaves fall off
4. forest biome that receives at least 200 cm of rainfall each year
5. dry temperate area, home to large hooved animals, usually migratory, and animals that feed on them
6. biome that receives less than 25 cm of rainfall each year
7. This biome covers 70% of the earth.
8. This biome includes lakes, ponds, swamps, and rivers.

Page 86: Biome Matchup
1. e
2. a
3. d
4. b
5. c
6. g
7. f

Page 87: My Favorite Biome
Answers will vary.

Page 88: Pollution
Answers will vary. ***Sample answers:***
1. Cars give off carbon monoxide as they burn gas.
2. Factories release wastewater into streams and lakes.
3. In cities, people honk horns, play loud music, and yell.

Page 89: What Am I Doing?
Answers will vary.

EARTH SCIENCE

Page 90: Earth Materials
1. *Lithosphere* describes the solid portion of Earth's crust.
2. *Litho* means rock.
3. Earth's hydrosphere is the surface water on Earth.
4. *Hydro* means water.
5. Earth's atmosphere is the gaseous layer that surrounds Earth.
6. *Atmos* means gas.

Page 91: My Earth
Diagrams will vary. ***Sample diagram:***

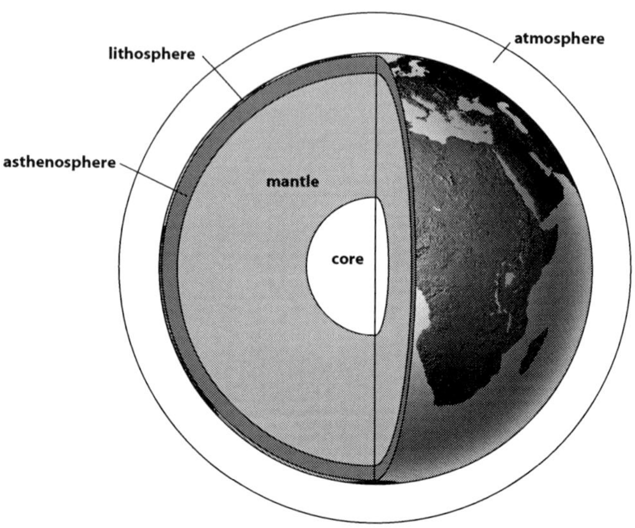

Page 92: Crystal Creations

1. cubic
2. tetragonal
3. triclinic
4. hexagonal
5. orthorhombic
6. triagonal

Descriptions will vary.

Page 93: Mineral Matchup

1. b	6. d
2. e	7. i
3. g	8. h
4. j	9. f
5. a	10. c

Page 94: Mineral Properties

1. the resistance of a mineral to scratching
2. the color of a powdered mineral
3. the shiny appearance of a mineral
4. breaks along a smooth, flat, parallel surface
5. the breaking of a mineral with little cleavage
6. the relative weight of a mineral compound compared to the weight of an equal volume of water

Page 95: Rocks On!

1. *metamorphic*—Rocks are changed because of exposure to heat and pressure.
   igneous—Rocks are formed by magma cooling.
   sedimentary—Rocks are formed by the breaking down of older rocks by weathering.
2. intrusive, extrusive; Pumice is light colored and not very massive. Its density is usually less than water. Obsidian, sometimes called volcanic glass, is dense and massive. It is usually black in color.
3. weathering, chemical precipitates, organic debris
4. temperature and pressure; temperature—heat increases as rock is pushed farther down in the Earth. Pressure also increases. Eventually, the rock is changed by the heat and pressure. For instance, limestone becomes marble.

Page 96: Resource Review

1. sunlight, water, wind
2. oil, coal, natural gas
3. Renewable resources can be replaced by nature. Nonrenewable resources cannot be replaced.
4. Nonrenewable resources will run out. There will be greater competition for renewable resources.

Page 97: Mineral Mining

1. T
2. F; They are called earths or rare earths.
3. T
4. T
5. Paragraphs will vary.

Page 98: Energy Resources

1. Answers will vary. ***Sample answers:*** oil, coal, natural gas
2. the remains of once living organisms
3. Answers will vary. ***Sample answers:***
   solar: comes from the sun
   wind: comes from wind
   hydroelectric: comes from water

Page 99: Conservation Concepts

Paragraphs will vary.

Page 100: How Old Is That Rock?

Answers will vary. ***Sample answer:***
The lowest layer of sedimentary rock is the oldest, the granite is next, and the top layer of sedimentary rock is the youngest.

Page 101: Fossil Finder

Answers will vary. ***Sample answers:***

1. You work with very old objects, usually fossils of living organisms.
2. Permineralization—Water seeps into bone and other hard body parts. As the body parts decay, minerals that came in with the water replace them in exactly the right shape, leaving a permanent copy.
   Preservation—Some organisms were trapped in material that kept oxygen from reaching them, such as amber or tar.
   Molds and casts—Shapes of hard body parts or footprints were made in quick drying mud or sediments, and the shape held.
3. A mold is an imprint left behind when an object falls into mud or sediments and decays. A cast is made when this impression later fills with mud or sediments, leaving behind a 3-D object.
4. Diagrams will vary.

Page 102: Past Perfect

1. Fossils represent life at various stages of Earth's existence.
2. a succession of rocks that represents a geological time scale
3. Diagrams will vary.

Page 103: Geologic Time

1. Cenozoic, Mesozoic, Paleozoic, Precambrian
2. oldest to most recent:
   Precambrian era
   Paleozoic era
   Mesozoic era
   Cenozoic era

3. Precambrian—soft-bodied animals, like jellyfish, and bacteria
   Paleozoic—animals with hard body parts, like trilobites
   Mesozoic—dinosaurs, first birds
   Cenozoic—large mammals, first human beings
4. Drawings will vary.

Page 104: Astrology Versus Astronomy

1. Astrologers believe the planets and Sun and Moon have influence on human lives. Astronomers are scientists who study phenomena in the night sky.
2. Astronomy. Astronomers study how stars shine, the orbits of the planets, and other things. Astrologers do not.

Page 105: Around We Go!

1. Diagram should show the sun, and the Earth, with the nighttime side in shadow.
2. Diagram should show the Earth tilt, and the Earth at the four positions relative to the Sun during the year.

Page 106: Mysterious Moon

1. b
2. d
3. a
4. e
5. c
6. Diagrams will vary.

Page 107: Phase Craze

Sample diagrams:

new moon

first crescent

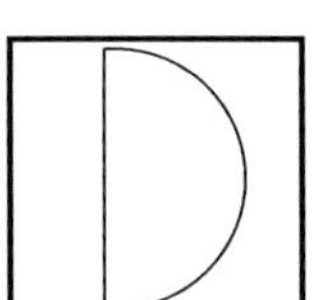
first quarter

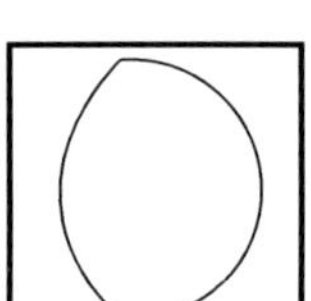
first gibbous

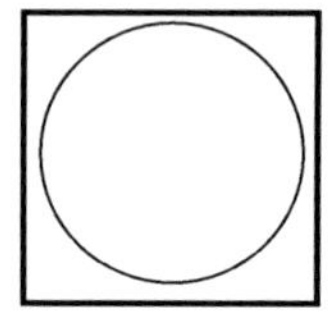
full moon

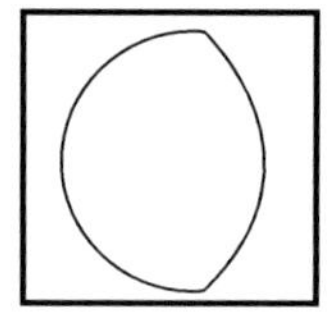
last gibbous

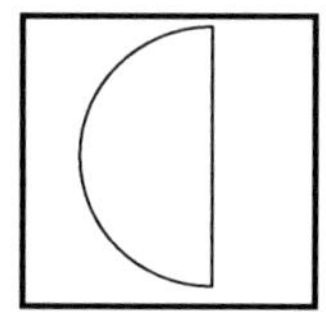
third quarter

last crescent

Page 108: Sun-sation!

1. solar flares
2. star
3. 150 million
4. Sunspots
5. photosphere
6. hydrogen and helium
7. corona

Page 109: Planet Power

1. Mercury
2. Mars
3. Earth
4. Venus
5. Mars
6. Mercury
7. Venus
8. Earth
9. Mercury
10. Venus

Page 110: The Outer Planets

Jupiter
1. fifth
2. 16
3. 13,003

Saturn
1. second
2. sixth
3. 17

Uranus
1. seventh
2. side
3. 15

Neptune
1. eighth
2. 2
3. Uranus

Pluto
1. ninth
2. one
3. Moon

Page 111: Our Solar System

Diagrams will vary.

Page 112: Planet Personality

1. h
2. d
3. b
4. g
5. e
6. f
7. c
8. a

Page 113: Planet Quest
Planets will vary.

Page 114: Don't Call Me Star
1. A meteor is in the atmosphere. Once it lands, it is a meteorite.
2. Comets are dirty snowballs made of dust and ice. Meteors and asteroids are chunks of rock and metal.
3. Asteroids are mainly in the asteroid belt, between Mars and Jupiter.
4. They are highly elliptical.

Page 115: Where Did It Go?
Diagrams will vary.

Page 116: Star Stuff
1. super giants
2. Proxima Centauri
3. magnitude
4. white dwarf
5. giants
6. quasars
7. radio telescope
8. dwarfs
9. black hole

Page 117: Life Cycle of a Star
1. nebula
2. protostar
3. star
4. red giant
5. black dwarf
6. white dwarf

Page 118: Constellations
1. Answers will vary.
2. The earth is rotating on its axis.
3. The earth travels around the Sun.

Page 119: Galaxies
1. spiral
   barred spiral
   elliptical
   irregular
2. Sketches of galaxies will vary.

Page 120: Telescopes
Diagrams will vary. Reflecting telescopes have mirrors; refracting telescopes have lenses. Both gather light and make stars seem brighter.

Page 121: The Space Shuttle
Paragraphs will vary. Some search terms include: Hubble Space Telescope, glider, reusable.

Page 122: The Ocean
1. a. continental shelf
   b. continental slope
   c. ocean floor
2. a. shelf starts where dry land ends
   b. slope begins where the shelf ends and slopes sharply
   c. floor begins where slope ends; it's flat and covers most of the ocean bottom
3. Diagrams will vary.

Page 123: Salts and Sediments
1. Water flows over land and collects salt; the water then flows into the ocean.
2. Diagrams will vary.
3. a. Rivers empty into oceans.
   b. Animals die and settle to the bottom.
   c. Plants die and settle to the bottom.
   d. Rock particles are carried to the ocean and settle to the bottom.

Page 124: Ocean Movements
1. current
2. trade winds
3. westerlies
4. Gulf Stream
5. California Current
6. upwelling
7. density
8. salinity; temperature
9. nutrients

Page 125: Waves
1. As wind moves over the water's surface, it pushes the water, causing waves.
2. Diagrams will vary.
3. Diagrams will vary.
4. a giant wave caused by an underwater earthquake

Page 126: The Story of Erosion
Stories will vary.

Page 127: How Is Soil Developed?
1. a. uppermost layer of soil
   b. middle layer of soil
   c. pieces of rocks not developed into soil
2. Diagrams will vary.
3. Answers will vary.

Page 128: Inside the Earth
Answers will vary. ***Sample answers:***
1. a. surface features, mountains, oceans, volcanoes
   b. plastic, convection
   c. iron and nickel, magnetic field comes from there
2. The sea floor is spreading.
3. eight large plates, and a number of smaller plates

Page 129: Earthquakes
1. c
2. b
3. e
4. d
5. a

6–8. Sentences will vary.

Page 130: Volcano!
1. Diagrams will vary.
2. cinder cone, composite, and shield; descriptions will vary.

Page 131: Way Up There!
1. a. the bending of rocks under steady pressure
   b. large blocks of rocks tilted over

c. forces below lift part of the crust
2. a large region of elevated land with a flat top
3. Large parts of the crust are bent upward.

Page 132: Up, Up, and Away!

Answers will vary. ***Sample answers:***

1. troposphere: where we live. All weather happens here; contains breathable atmosphere.
2. stratosphere: cold, not much oxygen. The ozone layer that protects us from ultraviolet radiation is here.
3. mesosphere: a cold, thin layer above the stratosphere. Meteors burn up here.
4. thermosphere: a hot layer. Contains the ionosphere where the aurora borealis happens. Space shuttles fly this high.
5. exosphere: the region bordering space

Page 133: Staying Cool Under Pressure?

1. The two factors are the amount of air above Earth's surface and the temperature.
2. Air particles prefer low pressure because there is less pressure.
3. A barometer measures pressure; answers will vary.
4. Some are reflected away; some are absorbed by the atmosphere; some are absorbed by the earth.
5. Diagrams will vary.

Page 134: Windy Weather

1. true
2. The wind pushes the plane.
3. false
4. When the ocean is cooler than the land, the air above it is cooled and moves toward the land. The reverse happens when the land is cooler than the ocean.
5. Answers will vary. ***Sample answer:*** Mountains push wind up, cooling it down and making it drop any moisture. The deserts of the southwestern United States are sheltered from the moisture because it is dropped over the mountains.

Page 135: Cloud Creations

Answers and sketches will vary. ***Sample answers:***
stratus—low-lying level of clouds, such as fog
cumulus—puffy fair weather clouds
cirrus—very high, thin, fair weather clouds

Page 136: Moisture Mind-Bender

1. rain
2. sleet
3. hail
4. snow

Explanations may vary.

Page 137: Fabulous Fronts

1. continental, maritime, polar, tropical
2. warm, cold, occluded
3. A warm front is formed, and extended periods of rain or snow occur.
4. A cold front is formed, and short periods of rain or snow occur.

Page 138: Why Forecast the Weather?

Paragraphs will vary.

Page 139: Tools of the Trade

1. b	3. a
2. c	4. d

Page 140: Weather Maps

Answers will vary. ***Sample answers:***
Surface maps show what is happening right now; forecast maps show what can be expected in the future.

SCIENCE AND TECHNOLOGY

Page 141: Natural Versus Synthetic

1. natural	6. synthetic
2. synthetic	7. natural
3. natural	8. synthetic
4. natural	9. natural
5. synthetic	

Page 142: Which Is Better?

Persuasive paragraphs will vary.

Page 143: Natural and Synthetic Do Battle

Comic strips will vary.

Page 144: Science Versus Technology

Answers will vary. ***Sample answers:***
Science: A biologist studies plants.
Technology: A biotechnologist studies ways to make plants more nutritious.

Page 145: How Can Technology Help?

Paragraphs will vary.

Page 146: Science and Technology Matchup

1. b	3. d
2. c	4. a

PERSONAL AND SOCIAL PERSPECTIVES

Page 147: Health Fact or Fiction?

1. a	4. a
2. a	5. b
3. b	

Page 148: The Magic Potion for Good Health?

Advertisements will vary.

Page 149: You Are What You Eat

1. about 2
2. 420
3. 2 g
4. 14 g
5. 12%
6. 4%
7. 86%

Page 150: Eating Healthy
Menus will vary.

Page 151: Is Your Eating Healthy?
Lists and paragraphs will vary.

Page 152: The Dangers of Drug Abuse
Posters will vary.

Page 153: Choices and Alcohol
Short stories will vary.

Page 154: Fighting Tobacco Use
Comic strips will vary.

Page 155: Renewable or Nonrenewable?
Answers will vary. ***Sample answers:***

1. Renewable resources never run out. Solar, wind, and hydropower are always available. Nonrenewable resources will eventually run out. We will run out of natural gas, oil, and coal eventually.
2. solar, wind, hydro
3. oil, natural gas, coal

Page 156: Energy Resources
Answers will vary. ***Sample answers:***

1. Fossil fuels are oil, natural gas, and coal. They were laid down during the time of the dinosaurs and before.
2. Solar energy is energy from the Sun. There are two kinds—solar panels, which convert the Sun's energy to electricity, and solar heating, which keeps houses warm and heats water.
3. Water energy is energy made when fast-moving water turns turbines in rivers and at dams. This is converted into electricity.
4. Geothermal energy is energy from the heat of the Earth. Geothermal energy plants are often found near geysers.
5. Nuclear energy is energy that is made when the nucleus of an atom is split or when two nuclei are fused.

Page 157: Living Resources
Answers will vary. ***Sample answers:***
Plants: Plants give off oxygen, provide food, and produce building materials such as wood.
Animals: Animals provide food, are used in hunting, can help people with disabilities, and are good companions.

Page 158: Identifying Populations
Answers will vary. ***Sample answer:***
butterflies, grass, raccoons, pine trees, squirrels

Page 159: Change Is Good
Answers and diagrams will vary.

Page 160: Technology Fun
Answers will vary.

Page 161: Good or Bad?
Paragraphs will vary.

THE HISTORY AND NATURE OF SCIENCE

Page 162: Zoologist

1. biology
2. animals
3. Answers may vary. ***Sample answers:*** zoo, laboratory, farm
4. Answers will vary.
5. Illustrations will vary.

Page 163: Botanist

1. biology
2. plants
3. Answers will vary. ***Sample answers:*** botanical gardens, laboratory, greenhouse, nursery
4. Answers will vary.
5. Illustrations will vary.

Page 164: Chemist

1. physical science
2. chemistry
3. Answers will vary. ***Sample answers:*** laboratory, university, hospital
4. Answers will vary.
5. Illustrations will vary.

Page 165: Physicist

1. physical science
2. physics; motion, light, sound
3. Answers will vary. ***Sample answers:*** laboratory, university, NASA
4. Answers will vary.
5. Illustrations will vary.

Page 166: Geologist

1. earth science
2. Earth's crust
3. Answers will vary. ***Sample answers:*** university, quarry, earthquake laboratory
4. Answers will vary.
5. Illustrations will vary.

Page 167: Astronomer

1. earth science
2. outer space
3. Answers will vary. ***Sample answers:*** observatory, university, NASA
4. Answers will vary.
5. Illustrations will vary.

Page 168: Marine Biologist

1. biology
2. marine life
3. Answers will vary. ***Sample answers:*** research ship, laboratory, university aquarium
4. Answers will vary.
5. Illustrations will vary.

Page 169: Oceanographer

1. earth science
2. the ocean
3. Answers will vary. ***Sample answers:*** research ship, laboratory, university
4. Answers will vary.
5. Illustrations will vary.

Page 170: Meteorologist

1. earth science
2. weather
3. Answers will vary. ***Sample answers:*** television weatherperson, weather station
4. Answers will vary.
5. Illustrations will vary.

Page 171: Isaac Newton

1. Newton's laws of motion
2. *First*—To move a body from rest, you must apply enough force to overcome inertia.
   Second—Force is equal to the mass of an object times its acceleration.
   Third—For every action, there is an equal and opposite reaction.

Page 172: Nicolaus Copernicus

1. Poland
2. *Born*—February 19, 1473
   Died—1543
3. He theorized that Earth and other planets revolve around the Sun.
4. No. Most people believed Earth was the center of the universe, and the Catholic Church also taught this.

Page 173: Thomas Edison

1. Milan, Ohio; February 11, 1847
2. Answers will vary. ***Sample answers:*** the light bulb, the phonograph, the moving picture camera
3. Answers will vary.

Page 174: Albert Einstein

1. Ulm, Germany; March 14, 1879
2. Answers will vary. ***Sample answer:*** Einstein's theory of special relativity states that everything in the universe is in motion. We figure our own motion relative to the motion of things around us. The only constant velocity is the speed of light, which is 300,000 km/s.
3. No; He was good in math and science, but terrible at language and history.

Page 175: Benjamin Franklin

1. He flew a kite in a lightning storm. Attached to the string was a metal key. As lightning struck the kite, electricity passed through the wet string to the key.
2. Illustrations will vary.

Page 176: Galileo Galilei

1. telescope
2. Jupiter's moons
3. Answers will vary. ***Sample answers:*** the rings of Saturn, the phases of the planet Venus
4. Physics. Galileo established the law of falling bodies, which is part of the law of inertia.

Page 177: Niels Bohr and Ernest Rutherford

1. the atom
2. Illustrations will vary.

Page 178: Wilhelm Roentgen

1. the X ray
2. Nobel Prize
3. her hand
4. X rays can look at the hard body parts of humans, such as our bones and teeth.
5. Answers will vary.

Page 179: Dmitry Mendeleyev

1. Russia
2. 1843–1907
3. He charted the known elements and predicted the existence of undiscovered ones.
4. gallium, scandium, germanium

Page 180: William Herschel

1. musician
2. Uranus
3. 1781
4. He used a homemade telescope.

WALCH PUBLISHING

Share Your Bright Ideas

We want to hear from you!

Your name____________________ Date__________

School name____________________

School address____________________

City__________ State____ Zip______ Phone number (____)______

Grade level(s) taught______ Subject area(s) taught__________

Where did you purchase this publication?__________

In what month do you purchase a majority of your supplements?__________

What moneys were used to purchase this product?

___School supplemental budget ___Federal/state funding ___Personal

Please "grade" this Walch publication in the following areas:

Quality of service you received when purchasing .... A B C D
Ease of use .... A B C D
Quality of content .... A B C D
Page layout .... A B C D
Organization of material .... A B C D
Suitability for grade level .... A B C D
Instructional value .... A B C D

COMMENTS:____________________

____

What specific supplemental materials would help you meet your current—or future—instructional needs?

____

Have you used other Walch publications? If so, which ones?__________

May we use your comments in upcoming communications? ___Yes ___No

Please **FAX** this completed form to **888-991-5755**, or mail it to

Customer Service, Walch Publishing, P. O. Box 658, Portland, ME 04104-0658

We will send you a **FREE GIFT** in appreciation of your feedback. **THANK YOU!**